Praise for *How the Canyon Became Grand*

"Stephen Pyne, historian and MacArthur Foundation Fellow, relates the Canyon's secrets in a book that interweaves geology, philosophy, and Darwinian biology. In *How the Canyon Became Grand* he also contends that when American civilization of the nineteenth century encountered the natural phenomenon of the Canyon, the land became place, and the place became an American icon . . . Pyne's book offers a superb intellectual and cultural journey."
—*Houston Chronicle*

"Pyne has written a genuinely mind-enlarging book."
—William McNeill, *The New York Review of Books*

"This book is a fascinating look at a fascinating work of nature."
—*Wisconsin State Journal*

"A fascinating, passionate history of one of the American West's last landscapes to be formally explored."
—*San Francisco Chronicle*

"Just as the Canyon demands physical rigor for those who would probe it deeply, *How the Canyon Became Grand* requires a certain intellectual sure-footedness through complex terrain. Both experiences are richly rewarding. For those who seek to understand the Canyon's evolution as cultural phenomenon, the book is a must."
—*Chicago Tribune*

"Pyne's genial nature and the manificence of his subject combine to make this tale one worth looking at."
—*Los Angeles Times*

"An energetic paean . . . leaves the impression that much of the rubbernecking people experience while floating the Canyon could be done, with equal excitement, in a gallery or library devoted to this singular piece of scenery."
—*The New York Times Book Review*

W9-AVZ-000

PENGUIN BOOKS

HOW THE CANYON BECAME GRAND

Stephen J. Pyne is a professor at Arizona State University, a MacArthur Foundation Fellow, and winner of the 1995 *Los Angeles Times* Robert Kirsch Award for body-of-work contribution to American letters. Acclaimed by *The New York Times Book Review* for his "major contribution to the literature of environmental studies," he is the author of the five-volume "Cycle of Fire," which includes *Fire in America* and *The Ice*. He also worked for fifteen seasons as a forest firefighter on the Grand Canyon's North Rim. He lives in Glendale, Arizona.

Books by Stephen J. Pyne

A Short History

Stephen J. Pyne

HOW

THE

CANYON

BECAME GRAND

PENGUIN BOOKS

PENGUIN BOOKS
Published by the Penguin Group
Penguin Putnam Inc., 375 Hudson Street,
New York, New York 10014, U.S.A.
Penguin Books Ltd, 27 Wrights Lane,
London W8 5TZ, England
Penguin Books Australia Ltd, Ringwood,
Victoria, Australia
Penguin Books Canada Ltd, 10 Alcorn Avenue,
Toronto, Ontario, Canada M4V 3B2
Penguin Books (N.Z.) Ltd, 182–190 Wairau Road,
Auckland 10, New Zealand

Penguin Books Ltd, Registered Offices:
Harmondsworth, Middlesex, England

First published in the United States of America by Viking Penguin,
a member of Penguin Putnam, Inc. 1998
Published in Penguin Books 1999

10 9 8 7 6 5 4 3 2 1

Illustration credits appear on page 201.

THE LIBRARY OF CONGRESS HAS CATALOGUED THE HARDCOVER AS FOLLOWS:
Pyne, Stephen J., date.
How the Canyon became Grand: a short history / Stephen J. Pyne.
p.   cm.
Includes bibliographical references and index.
ISBN 0-670-88110-4 (hc.)
ISBN 0 14 02.8056 1 (pbk.)
1. Grand Canyon (Ariz.)—History.   I. Title.
F788.P97   1998
979.1'32—dc21   98–20094

Printed in the United States of America
Set in Simoncini Garamond
Designed by Kathryn Parise
Topographic map pages viii–ix by L. Kubinyi

TO

William H. Goetzmann and John E. Sunder
who will know why

and Sonja
who has seen it happen

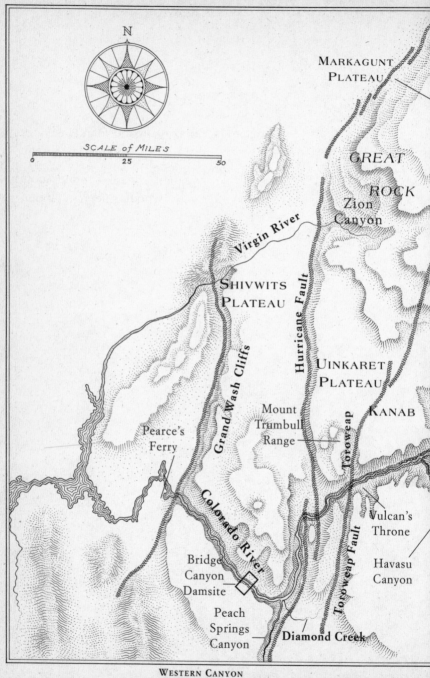

N

SCALE of MILES

0   25   50

MARKAGUNT
PLATEAU

GREAT

ROCK

Zion
Canyon

Virgin River

SHIVWITS
PLATEAU

Hurricane Fault

UINKARET
PLATEAU

KANAB

Grand Wash Cliffs

Mount
Trumbull
Range

Toroweap

Pearce's
Ferry

Vulcan's
Throne

Colorado River

Havasu
Canyon

Bridge
Canyon
Damsite

Toroweap Fault

Peach
Springs
Canyon

Diamond Creek

WESTERN CANYON

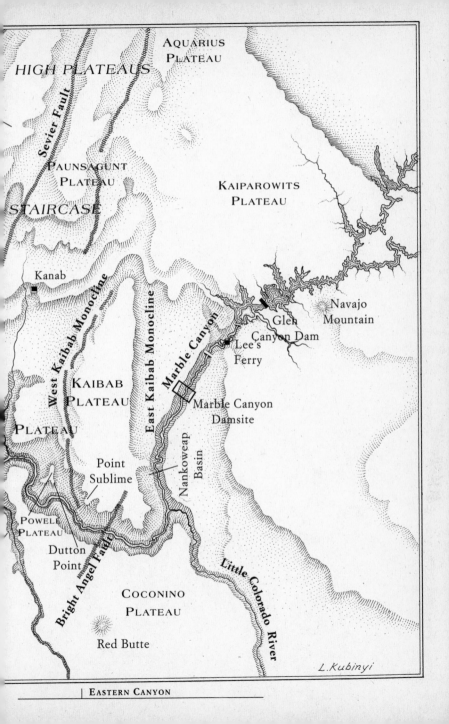

EASTERN CANYON

OVERLOOK:

THE VIEW FROM DUTTON POINT

Even for a landscape known for its uniqueness, this place is special. At Powell Plateau, an isolated mesa, the natural and human history of the Canyon distill into one compact monument. The mesa trails west like a pennant flapping in the winds of geologic time. A thick peninsula, concluding in Dutton Point, anchors that stony banner not only to the Canyon's gorge but also to the American civilization that claimed it. Stand at Dutton Point and see the features that make this landscape what it is.

All the geologic processes that have shaped the Canyon converge to sculpt a Canyon in miniature. The Muav fracture zone to the east separates the eastern and western Canyons and testifies to the forces that had pushed the plateaus to their great heights. An entrenched Colorado River arcing to the south and west shows the counterforce of erosion that turned a lithic hump into a hole. To the north there are more faults, spring-fed gorges, terraced strata, and the relentless recession of rock-wasting cliffs. Taken together they define the mesa's borders and document the events and processes large and small that have sculpted it. Even so, a perspective on the place is hard to maintain. By itself Powell Plateau stands taller than any peak east of the Rocky Mountains. Set within the Canyon, its dimensions are dwarfed and its distinctiveness compromised.

More important, the scene distills another Canyon, the one that most visitors actually see, a cultural Canyon, the Grand Canyon as a place with meaning. This landscape has been shaped by ideas, words, images, and experiences. Instead of faults, rivers, and mass wasting, the processes at work involved geopolitical upheavals and the swell of empires, the flow of art, literature, science, and philosophy, the chisel of mind against matter. These determined the shape of Canyon meaning. As they converged in place and time, they distinguished the Canyon from among hundreds of other, competing landscapes.

All this too is apparent from Dutton Point. Look southeast and see where Captain García López Cárdenas, equipped with odd baggage from the Renaissance, first peered over the South Rim searching for a river of empire. Look southwest and observe where Padre Francisco Tomás Garcés, carrying the continued burdens of the Counter-Reformation, entered the gorge through Cataract Canyon in a search for souls. Then came fur trappers and freebooters and agricultural colonizers. They peered and probed and, unable to discover much of value, let the gorges sink back into obscurity. Its size did not make the Canyon significant. Its physical geography alone could not establish meaning.

Look now about Powell Plateau itself. In one crowded throng of place-names, the mesa tells another story, a sudden outpouring of discovery and fevered interrogation as distinctive and separate as the mesa's physical presence. Its points all bear the names of Canyon explorers, from military commanders like Lieutenants Joseph C. Ives and George M. Wheeler to scientists like John Strong Newberry, John Wesley Powell, and Clarence Edward Dutton. One after another the major personalities of Canyon history had come to its brink. Here is where Powell directed Thomas Moran to the rim, and here the *Chasm of the Colorado* acquired its focal lens on the river. Here Dutton and William Henry Holmes came and looked east to the peninsula they named Point Sublime, from which they later immortalized the Canyon, in words and panoramas, as a Grand Ensemble. Here Uncle Jim Owens escorted Teddy Roosevelt to hunt mountain lions. At Dutton Point the Canyon's heroic age, figuratively and often literally, gathered to look and comment. Even amid the panoramic sweep of American society this overlook offered a special vision. Here a great civilization encountered a great natural phenomenon. Neither was the same afterward.

Why did they come and what did they see? At the heart of every Canyon overlook there lies the paradox that while indescribable, the scene is not incomprehensible. It has meaning, and that meaning depends less on the scene's physical geography than on the ideas through which it can be viewed and imagined. Those ideas are not something added to Canyon scenery, like a coat of paint, or taken from it, like a snapshot, any more than the river was something added to a prefabricated gorge. They have actively shaped the Canyon's meaning, without which it could hardly exist as a cultural spectacle.

The Grand Canyon was not so much revealed as created. More than once the Canyon was missed entirely or seen and dismissed. Then, with the suddenness of a summer storm, American society in the mid-nineteenth century mustered the capacity and the will to match its discovered opportunity and transformed land into place and place into symbol. The outcome was neither obvious nor inevitable. Popular instincts argued that river-dashed gorges were hazards, not adventures, and that immense chasms were geographic gaps, not gorgeous panoramas. A generation of intellectuals labored to instruct the public otherwise. They interpreted erosion-molded buttes as natural architecture and sculpture, as the coliseums, temples, and statuary of an inspired nature; they made folded blocks of crust into notebooks from the experimental laboratories of the earth; they rendered the etched strata of hard sandstone and friable shale into landscape frescoes and bas-reliefs of geologic antiquity. They made rocky rims beckon instead of frighten, shaped limestone-fractured peninsulas into focal points for a perspective of earth time instead of overlooks to voids, and allowed yellow, red, and mauve stone to radiate its own appreciated brilliance instead of merely betraying the absence of floral greenery. They elevated a ride down and through the gorge into the status of a national rite of passage. Until such transformations—until, in brief, the Canyon became a national monument as rich with cultivated meaning as a St. Peter's Basilica or a Crystal Palace—it was shunned. But once endowed with significance, like a rough diamond cut and placed under light, it dazzled, and America declared it beyond price.

The creation of Canyon meaning was as arduous and dramatic as the excavation of its great gorges. The process required a flow of ideas as vigorous as the Colorado River, and much as the scale of the Canyon depended on the height and mass of the excavated Kaibab Plateau, so the texture of understanding derived from the immense bulk of American civilization. The outcome was especially sweeping because for the Grand Canyon that interaction occurred between landscape and an educated elite—call it America's high culture—to large themes and high styles. No ancient folk resided between rims. No centuries-long tradition of utilization bound a peasantry to the place and shaped meaning with the slow patience of water dripping on a stone. Those who interpreted the Canyon would never live within it. They met it suddenly, and they could afford to see the gorges and mesas and sun-bleached terraces in idealistic and cosmopolitan terms.

Without that act of imagination the Canyon would likely have joined the throngs of forgotten and dismissed landscapes that litter the surfaces of the earth, a geographic freak, a landscape curiosity rather than a cultural oracle. But the improbable happened. Since the mid-nineteenth century, the Canyon has been explored and appreciated, and for the better part of a century it has been a central emblem, a profound truth, for many major American intellectuals and, through them, for American culture at large. In the end, the Grand Canyon cut a swath through the landscape of American history no less unique and grandiloquent than that which the Colorado River had excavated out of the Colorado Plateau. Indeed, the dramatic abruptness of the Canyon's brink matches the suddenness of that historic moment in which a Colorado canyon and American civilization met. How rim and river intersected remains a geologic mystery; how the place and its poets came together endures as something of an intellectual miracle.

No single vantage point captures all the Canyon. The panorama is too vast; the details of its evolution are too demanding. But the cultural Canyon did have its heroic age, during which it underwent a cultural chrysalis, and more than most the task of interpretation fell to—or rather was seized by—a curious polymath, Clarence Edward Dutton, captain of ordnance in the U.S. Army, geologist, raconteur, explorer, landscape critic, and author of that most comprehensive of Canyon books, *Tertiary History of the Grand Cañon District*. It was Dutton who argued that the Grand Canyon was a "great innovation in our modern ideas of scenery," that like all such innovations, it had to be understood before it could be appreciated, and cultivated before it could be understood. Then he showed how to do it.

So return to Powell Plateau and especially to that anchoring overlook that peers into the Canyon's core. Probably its scenic panorama is more comprehensive than majestic, the view more instructive than inspiring. It is a place not easily reached nor commonly visited nor readily appreciated. Its importance, in fact, lies less as a place than as a perspective; and for this it is unexcelled. Ultimately, Dutton's Point proposes a point of view, a positioning of the intellect to refract a whole culture, breaking like a sunset through prisms of sculpted rock and gilt-edged sky. So Clarence Dutton had argued; and so through Dutton can we refract the history of Canyon meaning. Stand at Dutton's Point and see how the Canyon became Grand.

CONTENTS

How the Canyon

Became Grand

TWO NEW WORLDS

I have heard rumors of visitors who were disappointed. The same people will be disappointed at the Day of Judgment. In fact, the Grand Canyon is a sort of landscape Day of Judgment. It is not a show place, a beauty spot, but a revelation.

—J. B. PRIESTLEY

From the top they could make out, apart from the canyon, some small boulders which seemed to be as high as a man. Those who went down and who reached them swore that they were taller than the great tower of Seville.

—PEDRO DE CASTEÑADA,
WITH THE CORONADO EXPEDITION

The gorges of the Colorado Plateau are remarkably elusive. Even so pronounced a landform as the Grand Canyon—prominent on satellite photos hundreds of miles aloft—is virtually invisible until one stands on the rim. It is possible to pass within a score of miles, or sometimes of meters, from that rim and never see the gorge, and more than one traveler has done just that. There is no measured transition: the plateau instantly ends; the canyon instantly begins. The rim is an edge, a weld between incommensurate landscapes. The river is more a barrier than a conduit. The Canyon suddenly is.

No impression of the place is more constantly invoked than the abruptness of its vision, a perspective almost wholly formed upon first view. The grand-manner tourist hotels erected along its rims exploited this fact, gave not a hint of the Canyon to motorists or train travelers until, as they passed through the foyer, the full spectacle burst upon them. A common query at the national park is the disarmingly perceptive, Where is the Grand Canyon? Few people have stumbled on to the Canyon; they had to set out with deliberation to find it. And to search it out, they needed an adequate reason.

The contrast with other landscapes is profound. Unlike a mountain's flank, the Canyon rim is defined with a geologic razor, and within its borders the Canyon is completely contained. A mountain may be seen from afar, its nature appreciated long before it is climbed, and for Western civilization the cultivation of mountain scenery had centuries behind it. So also rivers could be assessed from their tributaries, and traversed by them, one current feeding with measured flow into the other. So in other commanding landscapes known to Western civilization one feature led by gradation to another. But nothing led to the Canyon. It came as a phenomenon, an idea, and an aesthetic almost wholly without precedent. The Canyon suddenly was.

There was no evolved aesthetic or science for canyons as there was for mountains and waterfalls and other monuments of nature. For a defile on the scale of the Grand Canyon there was almost no prior preparation. The vision came—all of it, in all of its complexity and stunning uniqueness—as an instantaneous revelation. For those sixteenth-century Europeans who first encountered it, for whom the Renaissance was still aglow, the Grand Canyon was as unexpected an intellectual as it was a geographic enigma. They arrived more or less by accident.

But while the place remained, the civilization moved on. Those who came three hundred years later followed a scientific revolu-

tion in natural history, a Romantic revolution in art and consciousness, and a wave of democratic political revolutions. For them the
Canyon called, and the abruptness of Canyon scenery became a
tenet of its appreciation, part of an aesthetic canon of Canyon
mannerisms, and the abruptness of its rims, an illustration of the
erosive mechanics that had sculpted the region. That transformation in appreciation came, when it finally appeared, with a suddenness and completeness that seem peculiarly appropriate.

CANYON, FOUND AND LOST

No one knows what significance the Canyon held for the peoples
who came and went and for periods of decades or centuries lived
around the region over the ten thousand inhabited years before
Europeans arrived. Paleo hunters pursued bighorn sheep and
deer and left split-twig figurines in caves. In later times foragers
and farmers constructed stone dwellings along the rims and built
runoff terraces on Walhalla Plateau and grew beans, squash, and
maize around even inner Canyon springs and the Nankoweap and
Chuar deltas and laid out trails from rim to river.

What the place meant cognitively is unknown, or how each of
the handful or hundreds of peoples who saw it refracted the scene
through the prism of their aesthetic sensibilities or repositioned it
within their own moral geographies. Its magnitude alone demanded explanation and probably attracted some mythic significance commensurate with its physical dimensions and utilization.
Probably each people believed the space theirs, and themselves
chosen. The Hopi located the *sipapu,* the orifice through which
they emerged from the earth, near the junction of the Little Colorado with the Canyon. For the Havasupai, who farmed in the
well-watered Havasu Canyon, a tributary to the Colorado, and
hunted and foraged on the plateau's rim, the greater Canyon was
the border to where they lived.

But what other peoples believed about the site they kept to

themselves. Every society, after all, has its own sacred places. The Canyon became transnationally grand only after far-voyaging Europeans peered into it and then only after a much-metamorphosed European culture created a metaphoric matrix by which to interpret it.

Renaissance Europe encountered two new worlds, one of learning, another of geographic discovery. They were not always or necessarily fused. Scholars who constructed gorgeous *mappae mundi* with Jerusalem squarely in the center of the world had little in common with pilots who kept rutters and consulted empirically drawn portolan charts. That is why the Canyon was discovered quickly after the Great Voyages and why it was immediately forgotten.

The Canyon was, in fact, among the earliest of North America's natural wonders to be visited. Spanish conquistadors came to the South Rim in 1540, earlier by 138 years to Father Hennepin's sighting of Niagara Falls, by 167 years to John Colter's encounter with the Yellowstone, by almost 300 years to Joseph Reddeford Walker's discovery of Yosemite Valley. The Colorado River was identified and mapped long before the St. Lawrence, the Columbia, the Hudson, or even the Mississippi. Yet the Canyon was among the last of these wonders to be assimilated, much less celebrated. As far as Spain and the rest of Europe were concerned, the discovered Canyon quickly became a lost Canyon. While the sails of European expansion had swiftly reached the Colorado River, the Renaissance died on the voyages upstream and the overland *entradas* across its chromatic rocks.

The reason was in good measure due to the peculiar character of exploration in the sixteenth century. Beginning with the African coasting inspired by Henry the Navigator in the fifteenth century and concluding roughly with a revival of circumnavigations during the eighteenth century, well symbolized by the voyages of Captain James Cook, Europe had launched a great age of discovery, but one that was predominantly maritime. Leaving its inland seas, the Mediterranean and the Baltic, voyagers from Europe had sailed across the global ocean. The era's greatest discovery was the unity of the world sea; its grand gesture, a circumnavigation of that expanse; its outstanding achievement, a *mappa mundi* of the world's shorelines. Hammered for centuries from the Eurasian landmass, peninsular Europe had turned to the sea, and by stitching together, with the threads of its long-voyaging expeditions, previously segregated maritime regions, it produced a common quilt of the world ocean, the beginning of a truly global imperium. Newly discovered islands assisted that enterprise, but continents as often as not impeded it and provoked an endless search for straits, portages, or other passages around, over, or through them. For the Great Voyages the ship was the means, and the sea the end.

Europe's experiments in conquest, colonization, and commerce clung like ship's barnacles to the littoral of the world ocean or sought out offshore islands surrounded by sea moats. Everywhere outposts were founded on the coasts, and even the conquests of Mexico and Peru were preceded by the establishment of port cities, Veracruz and Lima. For the most part, penetration inland came by following major rivers that connected ports to the interior, or by crossing inland seas like the Great Lakes, or in a few exceptional instances by traversing overland.

Among the latter, however, were several *entradas* of epic proportion, and two of the greatest, those of Hernando de Soto and Francisco Vásquez de Coronado, set out in 1539 and 1540 respectively. Each had as its goal fabulous city-states, new Tenochitláns and Cuzcos, reputedly located somewhere in the interior, in Quivira. De Soto approached from Florida, Coronado from Mexico. Yet even the Coronado Expedition, at least in its conception, dared not abandon a maritime lifeline. While the main party probed northward into the American Southwest, another party under Melchior Díaz sought to rendezvous with ships commanded by Hernando de Alarcón moving up the Sea of Cortés to its rumored confluence with a great river. That rendezvous failed, but in an effort to reestablish such a waterway and during a time when "no other commissions pressed upon him," Pedro Álvarez de Tovar dispatched a party under García López de Cárdenas to investigate the rumor of a great river to the west of their winter camp at Zuñi. Such a river might be the same as that Alarcón was probing. It was worth investigating.[1]

With Indian guides from Tuzán to lead them, the Cárdenas party advanced to the Canyon's rim, probably near present-day Desert View. The site was cold and arid, covered with low-growing piñon and juniper. Proper perspective was impossible. The canyon looked like an outsize arroyo, the river little more than a fathom wide, boulders within the gorge no larger than a man. For three days, cold and thirsty, they probed for a way down. At last three members led by Captain Pablos de Melgosa attempted to scramble down at a place that "seemed less difficult." They returned that afternoon, having failed to reach the bottom, exclaiming that the Indians had been right, that the canyon was immense, that the river was broader than the Tagus and the perceived boulders taller "than the great tower of Seville." Of the chromatic view they said nothing. The river they could see but not reach. They had no means.[2]

With the wind biting, water scarce, and descent difficult, the Spaniards withdrew in disappointment. Intent on the discovery of civilizations, and with them gold to plunder and souls to convert, or on geographic discoveries that would lead to such conquests, the Spanish conquistador had little to say of the Grand Canyon. Only two chroniclers mention the foray; even Cárdenas says nothing in his *Relación*. Canyon geography proved to be little more than a false lead in the geopolitics of conquest, and an account of its exploration not much more than an aside in the narrative of an epic but politically futile trek. The Colorado River soon appeared on European maps by mid-century. It dominates North America on the Gastaldi map of 1546. But there is no indication of a great arroyo along the river's inland channel.

This indifference betrays something more than the steely soul of a conquistador. Coming three years before Copernicus published *De revolutionibus*, the symbolic prolegomenon to the scientific revolution, only twenty years after Magellan's fleet first circumnavigated the globe, and nearly thirty years before Mercator synthesized the known geography of the *terra nova orbis* with his famous projection, the Spanish had little context for the revelation of the Canyon. There were no scientists among the entourage, nor any artists; priests or personal secretaries doubled as chroniclers. Not for three hundred years would science even acquire its modern name. Granada, the last Moorish stronghold in Spain, had fallen to Spanish arms only in 1492, the year Columbus made landfall in the New World. Like Russia slowly sloughing off the Mongol yoke, Spain, in driving out the Moorish overlords, found itself curiously skewed to Western culture, an amalgam of Europe and North Africa as Russia was of Europe and Central Asia.

Where Spain had once led Europe's revival of learning, it now began to lag. The thirteenth century had experienced a renaissance of scholarship, culminating in the theological synthesis of Thomas Aquinas, that had derived in no small way from the recovery and translation of ancient texts. In this restoration Spain had been both center and conduit. It had prompted a reconciliation of Christianity with ancient learning. But the greater Renaissance that flourished in the fifteenth and sixteenth centuries passed over Spain without reaching, as it did elsewhere, the scientific revolution of the seventeenth.

Spain came hesitatingly to the experimental empiricism, the mechanical philosophy, and the secularism that increasingly informed Western thought. Renaissance yielded to a revival of religious orthodoxy, not modern science. By the time of Coronado's expedition Spain had become a stronghold of the Counter-Reformation; its scholars continued to meditate on the old texts and dismiss many of the new. Probably no European country was prepared to appreciate a phenomenon like the Canyon, but Spain was intellectually among the least receptive. While, in much of Europe, the Book of Nature joined Scripture as a testimony to the Creator, Spain proscribed it. That book proposed a different philosophy and demanded a different dialectic from that used to translate Arab-transmitted texts from antiquity. There were no Grand Canyons in the Aristotelian classics or the prophetic books of Holy Scripture.

Ironically, although Spain was perhaps the most advanced nation of Europe in its ability to mount expeditions and establish colonies, it was among the most retarded in its capacity to absorb its discoveries within the context of the new ideas and new sensibility that raged across the rest of Europe. Its interest in natural history focused largely on the question of native populations, a topic of supreme political, economic, and theological significance. Spanish rationalism was directed, in a dauntless rearguard action, toward the preservation of scholasticism. Five years after Captain Cárdenas halted on the Canyon rim, the Council of Trent began its counterattack against both Renaissance humanism and the practice of the new natural philosophy; eight years after Captain Melgosa clambered over the South Rim, there was born the greatest of Spanish philosophers, Francisco Suárez, a man whose twenty-six volumes consolidated Spanish metaphysics and theology on a thoroughly Thomistic basis.

In 1540 Spain demonstrated its imperial talents by expeditions of conquest, like those of Coronado and de Soto; by reforms in colonial administration, like the new Law of the Indies; and by the founding of the Jesuit order by a former soldier, Ignatius of Loyola. That the Cárdenas party could only liken Canyon features to those of Seville, the point of departure for Spain's overseas imperium, suggested both the power and the limitations of Spanish discovery. Its conquistadors were knights-errant, not savants. Yet Spain hardly stood alone. Decades were to pass before other nations began to penetrate the New World, and centuries were to unfold before European civilization could cope with what its explorers found.

The apparatus for valorizing such phenomena did not exist. Even had there been scientists with Cárdenas, there was hardly yet a cosmology suitable for interpreting a landscape as peculiar as the Canyon. The earth was believed to have commenced a few thousand years before, its great natural features shaped by the Noachian Flood. The invention of mathematical perspective was barely a century old. Cartographic projections, even those based on such methods, concentrated on the coastlines and oceans, not on the interiors. Had there been artists, they would have possessed few techniques on their palettes by which to convey the Canyon's immensity, awesomeness, and complex matrix of color and structure. Perspective had entered Spanish art only a handful of years before Coronado began his march; the conventions of modern landscape as formulated by Claude Lorrain were still a century in the future. Not for another 250 years would the calculation of longitude become more or less routine, would natural historians coin the word *geology,* would an educated elite begin to attach the word *sublime* to distinctive landscapes. With regard to learning sixteenth-century Spain was no worse off—was probably more advanced—than its European or Islamic rivals.

The celebration of natural monuments in and of themselves was alien to them all. Great arroyos held no value, not political, not economic, not intellectual, not aesthetic. None had any means by which to triangulate such a spectacle into a vision even remotely resembling the modern one. The Grand Canyon as a landscape fact attracted less attention than did geographic fables like Quivira, the Strait of Anian, or the Río Buenaventura. It was, after all, the fabulous but well-traveled yarns of Fray Marcos de Niza that had set the Coronado *entrada* into motion. That Spanish adventurers and missionaries made it to the Canyon rim is one of the marvels of Western history; that they failed to appreciate what they saw, one of its lesser mysteries. The Spanish mind was prepared to understand, and Spanish political economy prepared to assimilate, the discovery of Golden Cíbolas, not Grand Canyons.

For 256 years after Cárdenas no subsequent expedition

ventured to the Canyon's brink, and then the encounter was acci-
dental to the search for heathen souls. To Spain, the Canyon
remained a barrier, one of many along a daunting river that re-
fused to behave as a new Guadalquivir or Plata. The Grand
Canyon was an impenetrable tangle of *cañones,* mesas, and rapids,
uninhabited, inaccessible, peripheral, not a presence so much as
an absence, a place to be avoided. And so it was.

Still, the New Worlds of discovered geography did not remain in-
dependent from the new worlds of learning. Each complemented
(and challenged) the other, and together they questioned the
inherited worldview. The doubling in size of the known globe,
like the multiplication of ancient texts and critics, helped shatter
the authority of antiquity and antiquated methodology. The
breakdown of Ptolemaic astronomy, symbolized by the sixteenth-
century Copernican revolution, had its parallel in the disintegra-
tion of Ptolemaic geography under the blows of successive
voyages. The intellectual challenge to inherited cosmology involved both data and theory. Telescopes and explorers discredited
the presumed completeness of the synthesis recovered from the
ancients; mathematics, its theoretical organization; and the new
metaphysics of natural philosophy, its epistemology. Astronomers
discovered new moons unknown to the *Almagest.* Adventurers
unveiled immense new lands, populated with civilizations wholly
unsuspected by the *Geographia.* Humanist scholars found in
newly recovered texts from antiquity other geographies, unlike
that of Ptolomeaus and the Alexandrian school. Not only were
new data added, but over the centuries old errors were expunged.

The expanding imperium was intellectual as much as geopolitical, and exploration served both purposes. The Great Voyages became a model for empirical inquiry, a dramatic and intensely
practical manifestation of a *novum organon* in which the tested
reality of nature would substitute for the revelations of inherited
texts. Captains and their pilots to the New World had little to

learn from Herodotus, Lucretius, or Pliny the Elder. Their lives depended on understanding, as best they could, the shoals and currents and tides they actually encountered. So would modern scholarship. Francis Bacon captured that spirit nicely when he used as a frontispiece to his *Great Instauration* a picture of ships sailing beyond the Pillars of Hercules. In leaving the Mediterranean for America, European civilization was also venturing beyond antiquity's dominion of understanding.

Yet the two often ran in parallel, like a stream beside hills, the main currents of thought draining from and trimming the heights of land. The rush of the Renaissance passed. Overseas colonies claimed islands and clung to coasts, like seal rookeries. Exploration devolved into trade; piracy and poaching brought more rewards than sponsored discovery. Even the scientific revolution stalled. Having gathered momentum and inspiration over the seventeenth century from mathematics and experimentation, it reached a dazzling crescendo with Isaac Newton's synthesis of natural philosophy. That example of exalted reason radiated throughout European culture. The light of the new learning flooded into dark corners of superstition and ignorance; the Greater Enlightenment acquired its master metaphor. Then, after this wild burst of genius, exhaustion set in, an era of intellectual housekeeping. Scholars sought to illuminate inherited lore in the light of reason. They preferred to consolidate and codify existing knowledge and practices rather than seek out bold new philosophies. Metaphysical speculators and naive travelers became an object of ridicule, as in Jonathan Swift's *Gulliver's Travels*; squabbles over existing trade routes replaced the daring discovery of new ones; the importation and invention of new worlds gave way to dictionaries, and of imagined worlds to practical knowledge. Enlightenment scholars pored over windy texts and hypothetical maps through the sharpened vision that practical inventions like bifocals made possible.

Then the pace quickened again. By the mid-eighteenth century political rivalries and intellectual curiosity revived. Together empire and enlightenment began to stun geographic exploration out of its institutional fibrillation.

Second Age, Second Chance

The motivations spanned European culture. There was a renewal of extraterritorial expansion for which geopolitical rivalries were both cause and consequence. Russia commenced its imperial traverse of Eurasia; Britain and France inaugurated a new hundred years' war, this time sparring from India to Canada to the Antilles; the Netherlands poached Portuguese trade routes and even settlements from Brazil to the East Indies. Circumnavigation became the rage, and every nation aspiring to civilized status sponsored prolonged voyages of discovery. Every thrust brought European powers into conflict with one another as much as with indigenes. Coastal colonies in the Americas, Africa, and Australasia, drafting tens of millions of European émigrés, swept inland in a fantastic surge. By the early twentieth century Europe enjoyed political and economic hegemony over most of the planet. And from empire, exploration was never far removed.

Nor was learning. The Enlightenment spread first by applying new methods and perspectives to the existing canon. Scholars reworked inherited texts with the zeal of Renaissance humanists, this time not merely to translate but to rationalize. John Dryden rewrote Shakespeare to prune away fanciful language and excess. Alexander Pope translated Homer into heroic couplets. Everywhere savants sought to codify according to more modern criteria. Linnaeus proposed a *Systema naturae* by which to organize the flora and later the fauna of the world. Samuel Johnson assembled the first English dictionary. Denis Diderot oversaw the modern *Encyclopédie*, and Montesquieu, an annotated digest of the law. The age honored "practical" knowledge much as it sought a "plain" style.

Everywhere architects of Enlightenment tried to discover, or where they deemed necessary sought to impose, a reformed order based on their perception of Reason. But if the texts of the old authorities lay like broken idols, the concept of authority itself endured. The Enlightenment substituted and secularized; Nature replaced Scripture as a source of general authority; natural philosophy succeeded theology as the exemplar of knowledge. Newton's universe was as absolute in its structure as Aquinas's, though it removed dialectical logic for experimentation and allegory for mathematics. A full-blown relativity had to wait for modernism. The break with religion, however, at least helped quiet the era's politics. Europeans no longer slaughtered one another over minute differences between Newtonian and Leibnitzian calculus or among various models of gravity as they had over the mysteries of transsubstantiation and the practices of full- or partial-immersion baptism.

More important, the Enlightenment's momentum carried it beyond the status of a secular Renaissance. Its agenda had proposed a diffusion of learning—empirical, secular, mathematical, experimental where possible—that promised to penetrate every field of inquiry. It was inevitable that a period of consolidation and reconstruction should follow. But experimental science, unlike text-based scholasticism, could not long pause, nor long thrive with endless glosses; it needed more data, novel experiences, new worlds; and modern science (call it Newtonian after its greatest symbol) fueled the Enlightenment's engines. Natural science rested long enough to regroup, then it pushed on. It no more halted at putative frontiers than had Europe's adventurers. Its practitioners sought out new fields for study as eagerly as Clives and La Salles had new lands for claims and conquest. One dynamic, in fact, fed the other. Inevitably the imperial ambitions of Enlightenment learning fused with the imperial fervor of European expansion to yield both a new kind of explorer and a new era of exploration.

What William Goetzmann has termed the Second Great Age of Discovery, a revival of exploration that gathered force in the mid-eighteenth century, had many tributaries. As trade followed the flag, so did learning. The need to validate Newtonian mechanics, that exemplar of Enlightenment scholarship, could best be addressed by paired expeditions to the poles and equator, the famed surveys of Pierre-Louis Moreau de Maupertuis to Lapland and Charles-Marie de la Condamine to the Andes in the 1730s. The need to calibrate the Newtonian solar system, specifically to measure the distance between the earth and the sun, inspired a much-vaster enterprise in which in 1761 and again, on a larger scale, in 1769, astronomers surveyed the transit of Venus at sites around the world, from Philadelphia to Tomsk, from St. Helena to Tahiti. Meanwhile, Enlightenment science moved briskly from natural philosophy to natural history. By mid-century Carl von Linné had pioneered the natural history excursion, field-testing it on a series of traverses throughout Sweden and then, by means of his "apostles," twelve students he dispatched on foreign expeditions, throughout most of the European imperium, there to undertake an inventory of each region's biota.

Political and intellectual purposes soon converged. Probably nothing so characterized the transition as Captain James Cook's first voyage, a circumnavigation in which he traveled to Tahiti to survey the transit of Venus; carried one of Linnaeus's apostles, Daniel Solander; discovered New Zealand and eastern Australia, the sites of subsequent British colonization; and introduced Joseph Banks, one of the age's great naturalists and a founder and patron of the African Society and president of the Royal Society. "Every blockhead does that," Banks replied contemptuously of the social elite that traveled Europe; "my Grand Tour shall be one round the whole globe." Even enlightened despots sponsored analogous expeditions much as they did academies of science, as a symbol and instrument of the modernizing state. So Catherine II had sent Peter Pallas and a corps of savants to the land of Sibir (1768–74). From those studies came theories of importance to the cosmology of the earth—to geology, then a science without a name. Cook repeated his circumnagivation twice more before dying, a martyr to Enlightenment learning, on the third.[3]

But if Cook's circumnavigations effectively displayed the alliance of modern science with imperial ambitions, the final mapping of the world ocean's littoral, the extinction of the fabled Northwest Passage, and the abolition of a *Terra Australis* luxuriating in tropical splendor at the South Pole, the breakthrough came with the transfer of such methods to the continents. The Second Age would ride the wave of Europe's sweep over the earth's landmasses. Its grand gesture would be a traverse across a continent, a cross section of natural history. Its data, stories, and images would fill portfolios, stuff atlasses, line libraries with published personal narratives, stock the shelves of freshly endowed museums, and force the invention of new sciences like geology and of new genres of scholarship like anthropology. Moreover, much of what the First Age had discovered, the Second rediscovered. On the basis of his travels to South America, resulting in an encyclopedic survey of natural history, Alexander von Humboldt was popularly hailed as a "second Columbus."

That designation—applied to a foreign scientist, traversing the cores of Spain's New World empire—showed how far behind Spain had fallen as an intellectual power. Looking more imperious than imperial, Spain had for long decades stood outside the Enlightenment. Still equipped with the Inquisition and the Index, it remained wary of a secularizing rationalism, a transnational science, and an enthusiasm for overseas adventures that, it surmised, would likely come at its expense. Instead it closed its borders and fought the new learning with the same zeal it had the Reformation. And for so long as Spain remained beyond the reach of the Enlightenment, the Grand Canyon, nominally under the dominion of New Spain, would remain outside the Second Great Age of Discovery.

Eventually Spain, both peninsular and colonial, could no longer resist. Reform commenced with the ascendancy of Carlos III in 1759, effectively announcing a Spanish Enlightenment. The resulting reformation engulfed both empire and exploration. Northern New Spain, in particular, seemed threatened. For two centuries, valley by valley, tribe by tribe, the mission system had advanced northward, steadily bringing more land under the control of the crown. But in the eighteenth century the system faltered. By the 1760s, especially after a chastised France had been removed as an imperial contestant in 1763, New Spain undertook a general rehabilitation of its northern frontier. There were threats beyond the border from the British and the Russians, serious troubles within the *provincias* from the Apache, and vexing problems with the fabric of colonial administration itself. Provincial reformation proposed new routes of communication, projected a stronger cordon of presidios, mobilized the Royal Corps of Engineers, and substituted Franciscans for the Jesuits expelled in 1767.

It took longer, however, for field parties to gear up to Enlightenment standards. Pre-Bourbon Spain had tolerated the La Condamine expedition to Ecuador, though only after insisting that two reliable Spaniards accompany the corps to report on the suspected spies. It had also allowed one of the Linnaean apostles, Pehr Löfling, to visit Spain (1751) and accompany a foray to Venezuela (1754). With Carlos III, however, the crown endowed institutions of science—museums, botanical gardens, observatories—at which to train scholars to participate in the expeditions, particularly those within colonial possessions, that were becoming de rigueur for European powers. With the crown's sanction a Franco-Spanish team observed the transit of Venus from Baja California. A decade later the crown dispatched a botanical survey under Hipólito Ruiz and José Antonio Pavón to Peru and Chile, then granted permission to José Celestino Mutis to survey Nueva Granada (1783). The momentum provided by Carlos III even survived his death as Spain mounted two Cook-inspired enterprises:

the Royal Scientific Expedition to New Spain (1785–1800) and the Alejandro Malaspina Expedition to the Pacific (1789–94). But these were in style as much as setting light-years away from the *entradas* that passed along the Colorado canyons.

The first missed altogether, or avoided, the great gorge. While Juan Bautista de Anza pioneered an overland route from Monterey to Alta California, erecting missions and presidios, two priests, Father Silvestre Vélez de Escalante and Father Francisco Domínguez, accompanied by Captain Bernardo de Miera y Pacheco, a retired military engineer, trekked across much of the land that would become the Old Spanish Trail. The Escalante-Domínguez Expedition, it was hoped, would blaze an overland route between the Spanish settlements long established at Santa Fe with the new ones being developed in California. It was during this surge of Spanish interest in the region that the canyons of the Colorado River again entered the archives of Western civilization.

The party passed to the north of the Grand Canyon proper, made no attempt to inspect it, found a ford across the river (the Crossing of the Fathers), and reported on the character of the tribes and of the lands through which they passed. Though they failed in their larger objective, they successfully traversed much of the Colorado Plateau and, thanks to Miera's cartographic talents, gave a reasonably complete (though far from geodetically accurate) representation of the hydrography of the Colorado River system. They had no reason to visit the Canyon—had in fact every reason to shun it—and became as peripheral to its historiography as the Canyon was to their agenda.

Father Francisco Tomás Garcés, however, did see it. An old hand around the lands of Sonora, Garcés detached himself from the Anza Expedition, and under the direction of Indian guides pushed up the Colorado River to the Río Jabesua. This was the travertine-laden stream through Havasu Canyon, inhabited by the Havasupai Indians, a tribe linguistically related to the Yuman people who had served as guides.

From Havasu, Garcés evidently first encountered the western

Grand Canyon. Of the eastern Canyon to which Cárdenas had ridden, he witnessed nothing until he passed near the rim on a journey to the Hopi villages in June 1776. The Canyon he referred to as the Puerto de Bucareli, a gap by which the river passed through the sierra, and he referred to the river as the Río Colorado, sustaining the long-held Spanish belief that this river was continuous with that which empties into the Sea of Cortés. Because he brushed it only twice, and episodically, he missed completely or just misrepresented, as did Escalante and Domínguez, the great bend of the river—the central fact and mystery of its geography, the Colorado's east-west flow across the Kaibab Plateau— that had made the Canyon possible. But this lapse was of no consequence to the mission.

Of the setting, Garcés remarked only that he was "astonished at the roughness of this country, and at the barrier which nature had fixed" and, showing an inability to fathom the size of the phenomenon, thought "to all appearances would not seem to be very great the difficulty of reaching" the river. Only the size of the side canyons persuaded him otherwise. Evidently he was ignorant of the Coronado *relaciones*. Returning from the Hopi pueblos, Garcés again alluded to the gorge, this time as a "prison of cliffs and canyons." Probably his native guides steered him clear; his own cultural compass also pointed him elsewhere. He retired to missionary labors along the lower Colorado. Five years later he died during the Yuman uprising.[4]

Though Garcés visited the Canyon, it is obvious that it was not the object of his journeys. He traveled as part of a larger Spanish ambition to better integrate its frontier. Garcés, in particular, came to inspect the indigenous tribes along the Río Colorado, the most far-flung of which were the Havasupai, and when there were no longer any people, he left the river. He visited Havasu Canyon to size up its inhabitants and their agriculture; he cruised the rim of the Canyon only because it was, from Havasu, the most direct route to the well-known Hopi (Moqui) villages. The Canyon was an aside to his real purpose, and even then not one that especially interested him. Like Cárdenas, he saw nothing special about it except its superior ruggedness. Had the Havasupai not been present, he would not have made the trek at all. Hurrying directly to Moqui, he would have had no inkling that anything like the Canyon existed, even a league away, on the horizon.

The Canyon remained invisible. Between them the two expeditions had all but circumnavigated the Canyon, and while they continue to elicit admiration for the endurance and determination of their individual participants, they remain collectively irrelevant to the larger agenda of Enlightenment exploration. Even as their diaries, maps, and *relaciones* traveled to the authorities, their collective reconnaissance must have seemed like a heroic anachronism. But that was true, overall, for the Spanish Enlightenment itself. Spain's contributions to the Second Age of Discovery ended with explorer notes lost, unedited, and unpublished and its monumental expeditions in disarray. Once more geographic data were not broadcast as the common knowledge of transnational science, but hoarded, like other New World treasures, by the Spanish state. The Spanish Enlightenment imploded. What Carlos IV failed to resanction, the Napoleonic Wars terminated. Spain had dispatched Malaspina too late to the Cook-surveyed North Pacific and had failed to mount a natural history survey to places that might have rewarded the venture. It is as though a reinvigorated stream had veered out into a closed basin and dried into a salt playa.

The Canyon accounts more resemble the Jesuit *Relations* than the encyclopedic tomes characteristic of the new scholarship in natural history. Even as Escalante, Domínguez, and Garcés sent off their diaries and Miera his report and remarkable map, Peter Pallas was publishing the results of his expedition to Sibir and Captain James Cook's crew was returning, as melancholy heroes, from the last of the great voyager's circumnavigations. Pallas and Cook were the harbingers of a new style of discovery; Escalante, Domínguez, and Garcés resuscitated the old pattern of the *jornada* and the errant padre, the almost picaresque travels of the quixotic missionary guided from village to village by native Sancho Panzas.

Thus at a time when Pallas and the Linnaean apostles were promulgating a new mode of exploration, whose data virtually demanded a new science; when Georg Forster was publishing his *Voyage Round the World*, which helped established a new genre of literary travelogue and was to inflame the young Alexander von Humboldt; when William Bartram was completing his celebrated travels to the southeastern United States, the account of which was to inspire a generation of literary Romantics from Chateaubriand to Rousseau; when Britain's cultured elite considered a grand tour a necessary vehicle for eduction; when, in brief, new forms of art and science were evolving to describe nature's wonders, the Spanish explorers to the Grand Canyon are remarkable for their conservatism, their silence, their stubborn incuriosity about anything outside their prescribed agenda.

Their European and American counterparts were beginning to see even old scenes with new eyes. It is impossible to imagine Cook or Pallas, Linnaeus, Joseph Banks, or Daniel Solander before the scientific revolutions of the seventeenth and eighteenth centuries, or Forster and Bartram before the sensibilities of the later Enlightenment, yet it is easy so to imagine Escalante, Domínguez, and Garcés. There was no sense that what they saw required a new genre of literature or new ideas of physical science, or that the landscape they traversed represented a new metaphysic or aesthetic, or that, God willing, it was even worth a second look. Even that crowning achievement of Spanish exploration in the borderlands, Miera's map, shone because of its extensiveness, a triumph of practical empiricism, not because it was a new kind of map or a new way of looking at landscapes. It is as though Enlightenment physicians still read Galen or Newtonian physicists analyzed the solar system with the geometry of Eratosthenes.

As expeditions these sorties are almost completely interchangeable with Spanish surveys of a century (or two) before. As contributors to the history of ideas they are revealing in that they show a trend toward secularization and empiricism, but they neither derive from nor contribute to the revolution in scholarship exfoliating around them. Like the Spain that sponsored them, the motive behind the expeditions was conservative, defensive, and so was their interpretation. On the eve of the American Revolution, Spain sought to strengthen its colonial borders, and at the outset of an intellectual revolution, Spanish rationalism likewise strove to rebuild its frontiers, finding novelty somehow intrinsically threatening.

So in this latest round of reconnaissance, the Spanish carefully recorded oases of native tribes that they encountered, but not the natural splendors they saw. Certainly there was nothing in Garcés's account to give the impression that the Puerto de Bucareli was at all unique, much less that it required special language or new ideas, though far lesser scenes were beginning to inflame the imagination of European intellectuals. Even Thomas Jefferson was waxing wondrous about so prosaic a phenomenon as Virginia's Natural Bridge, and moody Britons on their grand tours swooned over fallen ruins and painters romanticized rustic Italy and literati hungry for the old and the exotic invented Highland epics like the saga of Ossian. Instead Spanish interest in the Colorado River, like Garcés himself, retired to its more accessible and populous lower stretches and there died.

When Alexander von Humboldt drew up his compendium of borderlands geography from Spanish (Mexican) sources in 1811, the Canyon appeared only as the Puerto de Bucareli, located at the junction of the Río Colorado and the Río Jaquesila (Little Colorado). The Colorado River was presented as running north and south, without the defining bend it makes through the Kaibab Plateau to form the Grand Canyon, thus eliminating the complex interaction of river and plateau that is the region's essence. Similarly, where Miera's map had been almost cluttered with ethnographic notes, Humboldt's meticulous compilation was sparse, employing the latest in cartographic technique and voiding the primary source of Spanish interest in the area, the Havasupai. Ironically it was Humboldt, a German geographer writing in French from Spanish sources, and Humboldt's *Map of the Kingdom of New Spain*, a part of his *Political Essay on the Kingdom of New Spain*, that finally introduced the region to the intellectual culture of Europe. The map became in turn the principal conduit for formal American knowledge about the region prior to the Mexican War.

The glamour of Spanish exploration came later, after European Romantics had transformed Spain into a tableau of the picturesque, and after similarly inclined historians like William Prescott had done for Old Mexico what Francis Parkman had done for New France. It was only later, after the Canyon had been publicized and its marvels extolled, that the Spanish contribution was recognized. Paradoxically it was not the Spanish accounts that helped make the Grand Canyon into an important emblem for Western civilization, but the Canyon, when once valorized, that gave meaning to the search for a cultural genealogy, that conveyed a value to Spanish travels not apparent at the time.

The encounter of Cárdenas with the Canyon in particular has become invested with a scholarship and significance out of all proportion to what seemed important to the *jornada*'s participants. Only two of the numerous *relaciones* bother to mention the episode at all, and only one of those expended more than two sentences in what the twentieth century has come to regard as a defining moment. Of all the events of that curious quest, its casual contact with the Canyon has become perhaps the most celebrated. For that Cárdenas and Coronado can thank the uninviting gorge they dismissed in their forlorn search for Quivira.

CONVERGENCE

The Canyon remained hidden until geopolitics met geopoetry; that synthesis required almost another century. By then imperial contests had transferred the region from an old and defensive Spain to a new and aggressive United States. Those same years, however, had witnessed a no less astonishing evolution and redefinition of cultural values that focused art, science, literature, philosophy, and nationalism on the acquired landscapes. The more majestic the scene, the more celebrated it became; the more singular, the more valued. That suited precisely the remote and peculiar canyons of the Colorado.

The threat to the Spanish colonies was real, but it came by revolution from within rather than invasion from without. From 1810 to 1821 Mexico, like much of New Spain, remained in revolt, eventually achieving independence. Where Spain had pursued a garrison policy along the frontier, sealing it off from inquisitive foreigners, Mexico quickly liberalized its policy of cross-border trade. In 1821, as the Ashley-Henry party pushed up the Missouri to inaugurate America's Rocky Mountain fur trade, other groups probed toward the Southwest, blazing the Santa Fe Trail. Fur trappers were prominent among the entrepreneurs of both expeditions; they turned Taos, especially, into a base of operations. Within a few years, if his *Personal Narrative* may be both believed and interpreted correctly, one such Taos trapper, James Ohio Pattie, passed around or through the Canyon.

During its heyday the fur trade revealed most of the scenic wonders of the Far West, though it often didn't identify them as such and left them with folk names like Colter's Hell and Brown's Hole. Parties of trappers exposed the Yellowstone, the Yosemite, the redwoods, the peaks and basins of the Rockies, the Great Salt Lake, and if they did not similarly dramatize the Grand Canyon, they apparently knew of its existence. The trapper had already become a stock character of American literature ever since John Filson had memorialized Daniel Boone in the late eighteenth century; by 1827 James Fenimore Cooper could make the Old

Trapper the center of adventure and moral drama in *The Prairie*. Often ghostwritten by professional literati, trapper accounts were surprisingly abundant, though frequently of doubtful reliability.

Pattie's *Personal Narrative* was one such piece of literature, and an especially problematic one at that. It was recorded by Timothy Flint, already renowned for biographies of Boone and others. Like all such narrators, Pattie (through Flint) claimed that he spoke only from plain facts, but again, like the rest, the story was embellished with common set pieces. The facts seem to be that Pattie was among the original group of trappers to the Southwest, that he traveled widely, that he visited the lower Colorado and Gila, and that he was imprisoned for a time by suspicious Mexican authorities.

Among his wanderings, however, he claimed in 1823 to have ascended the Colorado River to a place where "horrid mountains" shut in the river for nearly three hundred miles and prevented any descent. A rather nebulous, even feverish description gives only an impression of desolation. His allusion to the prisonlike feature of the gorge recalls Garcés and may reflect information Pattie picked up from the Mexicans during his captivity. His *Narrative*, moreover, proceeds as a series of burials and imprisonments, and his account of a similarly incarcerated river and canyon-buried trapper fits suspiciously well with that motif.[5]

Whether Pattie's *Personal Narrative* really describes the Grand Canyon or not, it does apparently give an accurate rendition of what from their travels Southwest trappers knew of the canyon country. They knew its general locale, that it was big, that it was desolate and impassable. Unlike his role with regard to so many other scenic highlights of the acquired West, the trapper for the most part worked to obscure the Canyon, not to reveal it. Antoine Leroux, for example, advised the Sitgreaves Expedition in 1851 not to continue along the drainage of the Little Colorado, for that would only bring it to a hopeless maze of rugged gorges. With other trappers as scouts, the Whipple Expedition a few years later passed by and named Red Butte, only a handful of miles from the

South Rim, but failed to approach the Canyon, even across terrain that offered not the slightest impediment to travel. The Coconino Plateau, which rises to the South Rim, was no more rugged than the swells of the Great Plains. Even the intrepid Jedediah Strong Smith, the Odysseus of the fur trade, missed the Canyon, following the Virgin River to the Colorado. His objective was the Old Spanish Trail, California, and beaver, not impassable gorges. The massacre of his party by Mohaves did nothing to improve the interest of the fur trade in this beaver-barren region.

In the end the American adventurers resembled nothing so much as secular padres, Franciscans of the fur trade, pursuing pelts rather than souls and as eager as missionaries for paths, not barriers. Certainly trappers said little about the landscape as a scenic or nationalist wonder. Yet there were stories—tales and folk fables not unlike those that circulated about Colter's Hell in the Yellowstone. When an Army expedition finally did proceed, with deliberation, to the region, its organizers already knew it by the name Big Cañon, likely a trapper sobriquet, and they knew "the accounts of one or two trappers, who professed to have seen the cañon, and propagated among their prairie companions incredible accounts of the stupendous character of the formation. . . ." Such reports "magnified" for the savants who followed "the marvellous story of Cárdenas," and if there were no trappers among the Ives Expedition to guide it to that scene, those yarns had the power to inspire to them, much as the tall tales of Fray Marcos de Niza had set Coronado on the path to Cibola. Before it became grand, the Canyon became fabulous.[6]

Meanwhile other forces were at work in the greater Southwest. In 1848 the Treaty of Guadalupe Hidalgo ceded most of the Spanish borderlands to the United States, and in 1853 the Gadsden Purchase added the lands of present-day Arizona south of the Gila River. The Canyon country found itself in the middle of a vast geopolitical realignment that would compel, eventually, an intellectual no less than a political assimilation. No longer was it part of an indigestible geography on a remote frontier of Spain or Mexico. It was instead a prominent natural feature of the major river of the Southwest, an American Nile, well within the heartland of Manifest Destiny. From the north came Mormon guides and colonists; from the east and south came traders up the Colorado and Army explorers searching for transportation routes. When the two processes finally met, they did so at the gateway to the western Grand Canyon.

Like their predecessors, Mormon frontiersmen were the tentacles of empire, and like them, the scouts mapped a practical geography, in this case of townsites and roads. Driven from Nauvoo, Illinois, in 1846, the Mormon hegira had resettled the Latter Day Saints on the shores of the Great Salt Lake. Joseph Smith, the martyred Mormon prophet, resembled nothing so much as the Emersonian poet-seer, arising like an avatar along the frontier, and the gathering of Mormon Zion in the Great Basin was a defining event in the geopolitics of western expansion. Powered by conversions and eager to expand Zion into the political state of Deseret, the Mormon state began to swell its territory by systematic colonization, especially to the south along the Wasatch Range and the fault valleys of the High Plateaus. Guides would investigate potential sites, and on the basis of their reports, whole communities could be "called" to settle the area. By 1851 towns, like Cedar City, hugging the western perimeter of the Colorado Plateau had been founded; by 1870, Kanab, at the foot of the plateau's Great Rock Staircase.

Perhaps the most celebrated of Mormon frontiersmen, Jacob Hamblin, had already circumambulated the Canyon region by 1862 and, like nearly every other visitor to the region, had found visits to the Hopi pueblos irresistible. But he avoided the Canyon proper, and so did Deseret. Mormon colonization veered to Lee's Ferry, where the Paria River joins the Colorado. Settlement then followed the Little Colorado south to a country of high meadows and forested mountains and eventually extended to the Mexican border. Another thrust fashioned a corridor to the sea along the Old Spanish Trail by way of Las Vegas and Los Angeles. Colonization grew around and incorporated the Canyon the way injured muscle might grow around an embedded arrowhead. What mattered were the Colorado's fords—Lee's Ferry, the Crossing of the Fathers, Pierce's Ferry—not gorges that could swallow Salt Lake City with hardly a trace.

The genius of Mormonism was for administration and religion: it created a new church and the machinery of empire, not a great literature, art, or science. New testaments to supplement the Bible, novel social institutions, a system of coinage, even an experimental alphabet—the cultural inventiveness was great, but it served an imagination intent on settling communities in an agriculturally forbidding landscape. No one concerned with irrigating crops in an arid land or searching out suitable timber, pasture, or townsites had the time or inclination to contemplate landscapes that offered little enough of any such things. The remembered sagas of settlement—the ordeal of the Muddy Mission, the intrepid trek to Hole-in-the-Rock—emphasized a social order that struggled against a formidable land. The land valorized their efforts by challenging the basis for their physical survival.

Nothing in their Scripture, not in the Old or New Testaments or in the plates buried by the prophet Moroni, said anything of revelatory canyons. Mormon imagination could interpret the deserts, mountains, and the salt lakes of the eastern Great Basin by analogy to the landscape of the Holy Land. Their chosen place thus complemented their creation story, the Exodus that had carried them across the Sinai of the Great American Desert. The promised land was a new Zion, a place to reclaim, not a picturesque Nature worthy of sublime awe and genteel contemplation or a secular text full of the lessons of natural philosophy and ready for explication by Gentile science. The Colorado canyons, in particular, were places to avoid or to ford.

While his acumen in political economics made Brigham Young one of the great captains of American industry, the probes he ordered to survey the landscape surrounding Deseret carried with them few of the instruments of high culture and shipped few such soundings back. Mormon settlers occupied such scenic marvels as Bryce Canyon, Canyonlands, Capitol Reef, Arches, and, with a rare gush of poetic enthusiasm, Zion Canyon. All framed riverine communities. Settlers along the Virgin River in particular could, with relative ease, enter and engage Zion's excavated valley and could allude to it as a kind of sandstone Yosemite. But of the Grand Canyon they said nothing. Within settled Deseret the Grand Canyon was as much wasteland as the salt playas of Sevier.

Yet even as Deseret swelled beyond the Great Basin, the Second Great Age of Discovery was overrunning the earth. The era's geopolitical sagas—the pursuit of a Passage to India, the rivalry between France and Britain over the South Seas, the westward Course of Empire across the North America, Britain and Russia's Great Game in Central Asia, Europe's unseemly Scramble for Africa—all were boulders rolled over imperial cliffs that scattered before them rockslides of exploring expeditions, sojourners, savants, and adventurers, a tumbling scree of the curious and the obsessed that constantly reshaped the topography of thought and the cultural contours of Western civilization.

In truth, geographic expansion was as much a cultural as a political process. The information the Second Age shipped back in bulging sea chests and personal narratives and steamer trunks of pressed plants, animal pelts, and artifacts of exotic indigenes forced a phase change in the Greater Enlightenment. Existing systems of scholarship could no longer cope with the rising flood of artifacts, impressions, weather recordings, logs, journals, specimens, barometric readings, drawings—the sheer volume of intellectual production that swelled over and ruptured the levees of inherited categories. By mid-century the floodcrest of the Second Age was reaching its high-water mark. Its emissaries were trekking across whole continents.

It was, accordingly, a great era of natural history for which Alexander von Humboldt serves as both promulgator and symbol. What Beethoven was to the music of the Romantic period, what Napoleon was to its politics, Humboldt was to its science. More than anyone else Humboldt transferred the apparatus of Cook-style exploration to the natural history of continents. Appropriately, Humboldt had been born in 1769, the year of Cook's transit of Venus voyage; he had taken inspiration from Georg Forster, the ethnographer with Cook's second voyage; and he was to dine with Thomas Jefferson the same month the Corps of Discovery under Captains Meriwether Lewis and William Clark departed from St. Louis on their epic journey across North America.

Lewis and Clark were beginning what Humboldt was already completing. His five-year expedition (1799–1804) to South America had pioneered a new genre of exploration. Even more, his exploits riveted the attention of Europe; his letters to his sister Caroline and brother Wilhelm, published as they arrived, made him a frontline journalist of discovery and a celebrity of geographic science. He had paddled up tropical rivers like the Orinoco, scaled peaks like Chimborazo, sketched the ruins of lost civilizations, experimented with electric eels, watched fruit-eating bats and carnivorous alligators, measured latitudes and mountain slopes, and obsessively collected, more than sixty thousand specimens in all. From landfall near Cumaná he wrote his brother Wilhelm that he and Aimé Bonpland, his traveling companion, were "like a couple of mad things. . . . Bonpland keeps telling me that he'll go out of his mind if the wonders don't cease soon." Here was the naturalist as Promethean hero. The contrast with the chroniclers of Coronado, with Escalante, Domínguez, and Garcés could hardly be greater.[7]

More than merely experiencing, however, Humboldt labored to bring the discovered new world of natural history into formal learning. His *Personal Narrative*, a partial rendering of his five-year travels through South America, expanded the model of his mentor, Forster, and established a literary genre for the scientific discovery of the exotic and the sublime. His fifty-four-volume *Voyage to the Equinoctial Regions of the New World* created an exemplar by which to appreciate and understand other newly discovered or rediscovered continents then entering European consciousness. In the process he helped invent modern geography and outfitted it with a methodology of comparative analysis and such organizing techniques as isothermals. His Thomistic *Cosmos*, first conceived in 1834, attempted to summarize, in vivid and popular language, both the new impressions and the expanded knowledge of nature freighted back by the science of his age. Here was the French *Encyclopédie* fused with German *Naturphilosophie*. Here incarnate was Tennyson's Ulysses, determined "to follow knowledge like a sinking star." Here, in the Humboldtean explorer, was the means by which to search out and peer over the Canyon rim.

Humboldt exercised special appeal to Americans and was a model explorer for a nation pushing into a western wilderness. An ardent democrat, he expressed vast hopes for the young Republic, and Young America responded enthusiastically. Emerson called him "a universal man, . . . one of those wonders of the world, like Aristotle." Walt Whitman, the self-styled "poet of the cosmos," attempted to synthesize with verse ensembles what Humboldt did with geographic science. Edgar Allan Poe dedicated his metaphysical prose poem *Eureka* to this giant of scientific discovery. The officers of the Army Corps of Topographic Engineers, the principal organ of government exploration, took him for a polestar; John Charles Frémont corresponded regularly. Louis Agassiz, professor at Harvard's Museum of Comparative Zoology, sought to synthesize North American natural history as Humboldt had South American. John Lloyd Stephens traveled to the Yucatán to rediscover the American Egypt that Humboldt had done so much to popularize; Federick Church, doyen of American landscape artists, trekked to South America to paint the grand scenes of Humboldt's adventures; and at the age of seventy-six, John Muir realized a lifelong ambition when he journeyed to the headwaters of the Amazon to find the flowers and trees Humboldt had described. When Americans rediscovered the Grand Canyon, they did so as Humboldteans.

They were almost too late. Humboldt had pioneered new intellectual ground, not only by proposing ideas but by inventing techniques with which to make sense of the Second Age's embarrassment of intellectual riches. Along with his contemporary Karl Ritter, he developed modern geography, integrating the thousands of newly discovered species, strata, and human artifacts through physical principles of geography; organizing them into ensembles of plants, suites of rocks, and communities of human settlement; analyzing them by systematic comparison. To the Great Chain of Being, Humboldt effectively added another dimension, cross-weaving its many parallel chains into a maplike grid. Yet the data continued to pour in, and the Humboldtean floodplain, however broad, could no longer hold it. By mid-century the Humboldtean explorer, like Humboldtean science, was already becoming a glorious anachronism.

Too much was revealed to be stuffed into the domain of geographic space alone, even into a cosmos like Humboldt's. Humboldt the geographer had tried to organize a progressively more complex universe through "laws" of space, while most metaphysically inclined thinkers were increasingly following Georg Wilhelm Friedrich Hegel's appeal to laws of time. More and more, intellectuals became natural *historians*. Behind that impulse, like a flood pushing against saturated levees, pressed the mounting evidence of exploration. The discovery of new lands was matched by the exhumation of landscapes lost in time. Fossils supplemented surface ruins; mastodons and dinosaurs, denizens of past epochs, added to the exotica of nature; ancient geologic empires were found in metamorphosed strata much as forgotten civilizations were dug out of desert sands and hacked from jungle vines. They could not be explained except by appeal to history.

The antiquity of the earth mattered. Whether the universe followed chance or design; whether humans had evolved from other life-forms or experienced a separate creation; whether time was a presence or a principle—all depended on how old the earth really was and how its history was organized. The Humboldtean became a Darwinian, broadly defined. With eerie timing, the aged Humboldt, born the year Captain Cook set out to measure the transit of Venus, died with *Cosmos* still—inevitably—incomplete in 1859, the year Charles Darwin published *On the Origin of Species*, the outgrowth of his own five-year voyage of discovery.

So they came together: the Greater Enlightenment, the Second Age, the colonization of continental interiors, the unprecedented terrains of America's Far West. The compound exploded Western civilization's horizons of geography, history, and perception. Nowhere were its shock waves felt more powerfully than in the United States, a self-consciously new nation as eager for a past as for a future, for which nature often substituted for culture and the westward migration of which coincided precisely with the broader parameters of European expansion.

Natural history and national history proceeded in sync, a cultural fugue to Manifest Destiny. A new people shaped by colonization, democratic revolution, and industrialization found particular confidence in the idea of progress, for which nature's evolution seemed to furnish both precedent and precondition. The national creation story commemorated the encounter of Old World civilization with New World wilderness; the contest for empire and its westward expansion was the national saga, the American *Aeneid*; history interpenetrated geography to make this a march into the future. The frontiersman, forged along this border, close to nature, became the putative wellspring of American virtue. The explorer served as guide, an almost Moses-like figure.

The national epic found its monuments, as often as not, in the American landscape. Nature, which America had in abundance, replaced the built environments that it lacked. Niagara Falls mocked the fall line of Europe's overshot water mills and even more the contrived fountains of Versailles or Bernini's Fountain of the Four Rivers in the Piazza Navona. The Rockies dwarfed the Alps. The majestic Mississippi lorded over a quaint Rhine. The Great Lakes reduced Europe's waterscapes to ponds. Redwoods defied Europe's hoary oaks, grizzly bears its domesticated farmyard fauna, Yellowstone's geysers the lapdog hot springs of European spas. The natural, the big, the distinctive—all challenged the artifice of ancient and aristocratic societies, while arguing strenuously for a republic of native, once-and-future virtue.

The unveiled spectacles begged to be seen and drew artists like bees to clover. Some inventoried the lands with the zeal of Linnaean apostles. Mark Catesby tabulated flora and fauna; John Jacob Audubon recorded birds and quadrupeds; George Catlin feverishly documented (and idealized) the indigenous peoples before they passed away forever. But the same fervor soon applied to landscapes, and landscape quickly became America's dominant art form, certainly its most popular. The moral didacticism that had previously been invested in grand-manner history paintings now infused the scenes of natural history. A genre pioneered by isolated artists like Thomas Cole rapidly blossomed into regional and national stature like the Hudson River school and went international. Frederick Church, the outstanding American artist of mid-century, carried the landscape beyond Niagara Falls and the Adirondacks—American wilds, removed from rural gentility—to the South American vistas popularized by Humboldt and even to Greenland. Germany's Düsseldorf school formalized practice with theory and elevated landscape into the grandest of arts, the opera of painting. Americans like Albert Bierstadt compounded both, studying at Düsseldorf and then journeying to the Rockies and the Sierra Nevada. Artists thus sought out the remote, the sublime, the monumental, and the distinctive and often exploring expeditions that could take them to such places.

Thus the historical tributaries all converged, like the three forks of the Missouri, to make a pathway into the interior. Expansion had acquired new, seemingly vacant lands; modern science turned its attention precisely to such landscapes, pushing natural history into exponential growth, reinventing sciences like geography and biology, spawning new sciences like geology and anthropology by rapid fission; the arts celebrated landscape as its highest expression; literature thrived on travelogues and personal narratives of adventuring expeditions; nationalism demanded new monuments and found them in the landscape-sculpted Far West; Romanticism in the arts and philosophy celebrated the natural over the artificial and the new over the inherited. All these currents converged into the exploring expedition, and as the American empire surged westward, the federal government endowed, as its official organ of discovery, the Army Corps of Topographical Engineers. It is estimated that the U.S. government invested as much as 25 percent of its antebellum budgets in just such exploration.

The corps conducted a series of expeditions that far surpassed its political charge to record transportation routes and survey boundaries. It was in fact an institution of the Second Age, and it oversaw a "great reconnaissance," as Edward Wallace and William Goetzmann have aptly termed it, that was both a consequence of and a stimulant to the revolutions in science and sensibility that gave the age its dynamism and color. Educated at West Point, a professional engineer as well as a soldier, conversant with the scientific and artistic luminaries of the day, often with European travel as well as western exploration behind him, the topographic engineer "considered himself by schooling and profession one of a company of savants." He became an agent of cultural no less than political expansion. When the corps was authorized in 1838, it postdated by only two years Ralph Waldo Emerson's transcendentalist manifesto *Nature* and Thomas Cole's inaugural landscape *The Oxbow*. When Lieutenant G. K. Warren summarized the corps's collective discoveries in cartographic form, the map preceded by two years the first edition of Charles Darwin's *Origin of Species*. It was an era that looked to nature for inspiration as well as information; the corps was a prime vehicle of that inquiry.[8]

The corps's exploration of the canyon country spanned the 1850s. There came, first, the Sitgreaves Expedition (1851), searching for wagon roads across northern Arizona; then the Whipple Expedition (1853), plotting out a potential route for a transcontinental railroad; then Beale's Expedition (1858), improbably crossing the Great American Desert by camel; and finally, a culmination, the supreme achievement of two decades of dedicated exploration, the Ives Expedition (1857–58), hopefully paddling up the Colorado River by steamboat and recording, for the first time since Garcés, an encounter with the Canyon.

The stimulant was a collision of political ambitions that pitted Mormon Deseret against the United States at a time when Bleeding Kansas and sectional divisiveness in general threatened the integrity of the Republic altogether. The so-called Utah War is best known for the Mountain Meadows Massacre and the occupation of Salt Lake City by federal troops under Albert Sidney Johnston. But the Ives Expedition, nominally a search for transportation and supply routes to Utah, an uncanny echo of the Coronado and Alarcón missions, may be its most enduring legacy. The final result, not fully realized until the Civil War ended sectional secessionism, was the integration of the region into the United States, not into a quasi-autonomous Deseret, and into the cultural—indeed moral—geography of the Second Age. Mormon exploration was intensely practical, a search for sites that could replicate and sustain the plat of Zion that formed the basis for communities called to colonize. The Canyon held nothing of value, was in fact repugnant, to such purposes. The corps brought with it different ambitions. It was eager for the exotic, the spectacular, the scientifically instructive, the curiously picturesque.

Consequently, although Mormon scouts probably preceded the Army and while some scouts certainly shadowed Army parties as they worked through river and gorges, it was the corps, with its cartographers, naturalists, and foreign eccentrics, that penetrated the western Grand Canyon, that began the assimilation of the Canyon into American political institutions and intellectual traditions, that allowed the rapids of the Colorado River to enter the mainstream of American ideas. The Ives Expedition commemorated two encounters, that between Deseret and the Army of the United States and that between the Second Great Age of Discovery and the Grand Canyon.

Stream and hill, process and place, so long in parallel, closed. American civilization in all its snarled currents and swirling exuberance converged on a landscape uniquely its own. The revelation of the Canyon—the Canyon itself as a kind of revealed knowledge—finally followed when that cultural river deepened and turned and the mountain rose to meet it. The two new worlds merged at last.

RIM AND RIVER

Nowhere on the earth's surface, so far as we know, are the secrets of its structure revealed as here.

—JOHN STRONG NEWBERRY

If any of these stupendous creations had been planted upon the
plains of central Europe, it would have influenced modern art
as profoundly as Fusiyama has influenced the decorative art of
Japan.

—CLARENCE EDWARD DUTTON

Before 1857 the Canyon was an incidental landform, concealed
amid scores of exotic western scenes, no more distinguished
than the ancient shorelines of the Great Basin or the glaciated
summit of Mount Shasta. Hell's Canyon on the Snake River was
deeper. The channeled scablands of the Columbia were more sav-
agely eroded. The Colorado River itself had excavated a dozen
other canyons, through strata more singular and monumental.
The Far West abounded in geographic marvels. Zion Canyon,
Arches, and Monument Valley—all on the Colorado Plateau—
exhibited land sculpturing on a more humanly impressive scale.

But the Canyon gradually transcended them all. From 1869 to

1882 it went from the status of a legendary giant suck to the subject of two classic works of American letters, from a place shunned even by professional pathfinders to one sought out by scholars and tourists with evangelical zeal and to which, at considerable inconvenience, the 1891 International Geological Congress would be directed. A peripheral landscape without cultural precedent—a scene as alien to Western civilization as the plains of Mars or the craters of Mercury—had seized the center and become an exemplar of geology, an epitome of historicism, a talisman of landscape art, and an icon of American nationalism. In 1903 Teddy Roosevelt, then president of the United States, rode a train along a specially constructed spur track to the opulent El Tovar Hotel on the South Rim and proclaimed to reporters of the *New York Sun* that the landscape before them was one of the "great sights every American should see."[1]

In roughly forty years the Canyon had become Grand.

LONELY AND MAJESTIC WAY: BIG CAÑON

The Colorado Exploring Expedition (1857–58) under Lieutenant Joseph Christmas Ives was among the last of the antebellum surveys directed by the Army Corps of Topographic Engineers and was in many respects its apex. It had all of the strengths of the Humboldtean mode, and most of its weaknesses. Almost immediately upon its completion America's exploratory and imperial impulses experienced a profound redefinition. *Report upon the Colorado River of the West*, published in 1861, appeared at the outbreak of the Civil War and two years before the formal dissolution of the corps. No less significantly, it occurred amid a major reorientation in natural science and social theory, aptly timed with the debate over Darwin's theory of evolution by natural selection. Symbolically, the vehicle for Ives's expedition, the steamboat *Ex-*

plorer, slammed into submerged rocks at the entry to Black Canyon while its crew were "eagerly gazing into the mysterious depths beyond. . . ." One might say the same for the Romantic haze through which latter-day Humboldtean exploration tried to proceed.[2]

The expedition's specific charge was to determine the limit of navigation to the Colorado River. But its ranks were staffed with a sampling of the era's educated elite, not only the military engineer and explorer Ives but a physician-naturalist, John Strong Newberry; a German artist, H. B. Möllhausen; and a Prussian-born cartographer, Baron F. W. von Egloffstein, all veterans of Army exploration in the Far West, all fellow travelers of the Second Age. More than an experiment in hydrography, the expedition proposed a full-bore survey of natural history. With the *Explorer* beached for repairs, Ives satisfied their formal charge by piloting a skiff up Black Canyon and, with the water low, the gorge deep, and the rapids frequent, determined that here indeed was the Ultima Thule of river transport. To meet the expedition's second, grander purpose, he divided his command, sending half back downstream and organizing the remainder, including his scientists and artists, and marched overland to the east.

With native guides, hence along roughly the traditional routes of human traffic, Ives recapitulated (and elaborated) Garcés's itinerary. Ives's principal innovation, the one on which the expedition's reputation was to rest, was to descend down Peach Springs Canyon to the inner gorge of the western Grand Canyon, where Diamond Creek joins the Colorado. It was an easy route, down a tributary canyon wide enough to accommodate a wagon had Ives chosen to force one. There, at the junction, the corps of discovery remained for several days, contemplating, sketching, musing, until finally deserted by their guides. The party then resurfaced, moved east, and reentered the Canyon at Cataract Canyon (Havasu) before again retiring, this time for good. The expedition followed existing trails and corps-blazed wagon roads to Fort Defiance, New Mexico, where it mustered out.

The expedition missed as much as it found. Of the eastern Grand Canyon or of the confluence between the Colorado and the Little Colorado, it saw nothing, and its published map reflected this ignorance. It extended the Big Cañon westward to include Black Canyon and confused the relationship between mainstream and tributary, making the Little Colorado River into the main branch, and it diverted the Colorado proper into Kanab Canyon to the inaccessible north. For all the topographic care invested in its cartography, the expedition's hydrography was worse than Miera's nearly two centuries before. But its descent down Diamond Creek marked the first journey from rim to river, and with that passage the expedition carted the Second Great Age of Discovery face-to-face with perhaps the most fantastic of the innumerable landscapes that era would confront. Its collective contribution far exceeded the perspective of a lieutenant enamored with the prospects of literary fame and harassed by unreliable guides and perpetual shortages of drinking water.

Joseph Ives was a Yale man, a West Pointer, and a veteran of the Whipple Expedition that had traversed northern Arizona in 1853 as part of the Pacific Railroad surveys. He knew the accounts of Cárdenas and Escalante and knew, probably from trapper stories, about the Big Cañon somewhere up the Colorado. He could envision himself as a Humboldtean knight-errant and could appreciate the exploration of the mysterious Big Cañon as a quest. "He talked of the Colorado expedition," as even one critic admitted, "as 'the event of his life,' destined to make fame for his children." In that judgment he was right. So was his assessment that as military transport routes the landscape and river were useless, and as sites for agriculture or settlement, hopeless. Unfortunately in other judgments he erred, sometimes spectacularly.[3]

The central narrative for the *Report* was Ives's. Most of his prose was charming rather than overwrought, and his eye for the picturesque rested most often on the natives, like the irascible Mohave chief Ireteba, who accompanied the party for long weeks. Yet the idea of the sublime—more convention than concept, no longer infusing awe with terror—was never far from his pen. He surveyed the scenes with "wondering delight," citing particularly those prolific examples of its most picturesque forms: its "gigantic chasms," like a "vast ruin"; "isolated mountains," like natural pyramids; fissures "so profound that the eye cannot penetrate their depths"; spires "that seem to be tottering upon their bases," rising thousands of feet like Egyptian obelisks. At times it seemed that the Colorado Exploring Expedition had metamorphosed into the Institut d'Égypte sent to the Nile by Napoleon. Within the Canyon a Gothic gloom pervaded Ives's field of vision, as though the passage down Diamond Creek had been an Odyssean descent to the underworld.[4]

Then he indulged in a rhetorical flourish that made his narrative more or less notorious in Canyon history. "The region is, of course, altogether valueless," he concluded. "It can be approached only from the south, and after entering it there is nothing to do but leave. Ours has been the first, and will doubtless be the last, party of whites to visit this profitless locality. It seems intended by nature that the Colorado River, along the greater portion of its lonely and majestic way, shall be forever unvisited and undisturbed." Ives meant that as a kind of compliment, and it was a fine expression of literary melancholy from an era of Romantic discovery whose agents had rediscovered the Alps and climbed the Andes, probed the Khyber and Bolan passes, uncovered ruins from the jungles of the Yucatán and stood, echoing Shelly's Ozymandias, as they overlooked civilizations submerged by desert sands. At the Big Cañon, nature itself lay in sublime ruin.[5]

But in the light of what was to follow, and even in the context of what his colleagues on the expedition were discovering, Ives could not have been more wrong. The Big Cañon was not a sphinx, obscuring in gloomy veils and gaping fissures nature's past, but a revelation, exposing the lost secrets of natural history. Melancholic, perhaps, but undoubtedly sublime. While Ives struggled to convey both sentiments, the very aesthetic and literary conventions that enhanced his appreciation also constricted his judgment. He could not envision a meaning beyond grotesque sculpturings, Gothic atmospherics, and picturesque prose.

Curiously the breakdown began at Black Canyon. Once they were beyond the head of navigation, prose and illustrations lost a sense of reality. Without having to measure, map, and test the river for utility as a transportation route, the Romantic imagination roamed unchecked by the imperatives of practical engineering. Almost everyone, from Ives down, indulged in it to some degree. Their points of reference were those predecessors who had blazed routes to the south, trails they left reluctantly and to which they eagerly returned. The head of navigation was also, it seems, the limit of critical perspective. After the expedition Ives's judgment seemingly worsened, to the further declension of his reputation. The ambitious lad from New Haven married into southern society and the Confederacy, served as aide-de-camp to Jefferson Davis, and was eventually buried in Oxford, Mississippi.

A similar intellectual schizophrenia afflicted the expedition's two artists, H. B. Möllhausen and Baron F. W. von Egloffstein, both Germans, both Humboldteans, and both veterans of the Pacific Railroad surveys, and both compromised between the landscape they saw and the one they were equipped (and avid) to express. Between them they introduced the Canyon to Western art. Their sketched and painted landscapes were visual cognates to Ives's Romanticized prose, and their judgment echoed his.

This was Heinrich Balduin Möllhausen's third expedition to the American West and his second to the Colorado River. In 1851 he had joined the excursion of Duke Paul of Württemberg to the Rocky Mountains, an adventure in Humboldtean natural history.

From 1853 to 1854 he served the Whipple Expedition across northern Arizona in the capacity of topographer. There he met Ives, another member. A return to Germany introduced him to the illustrious Humboldt himself. He proceeded to marry the daughter of the old explorer's personal secretary and returned with letters of introduction, a personal benediction, from the Second Age's éminence grise. "Mr. Möllhausen," Humboldt concluded, "dreams of nothing but of the happiness to be attached once more to an American expedition." His aspirations and Ives's coincided. The lieutenant hired him as "artist and collector in natural history."[6]

Möllhausen kept a journal and, as the expedition's official artist, sketched a visual log of what the party saw: flora, fauna, natives, expedition members, landscapes, above all the river, which Möllhausen assessed artistically as Ives did hydrographically. His personal narrative Möllhausen published in German, *Reisen in de Felsengebirge Nord Amerikas* [*Journey to the Rocky Mountains of North America*]; his sketches, suitably engraved, entered into Ives's official *Report* and, reworked into watercolors, Möllhausen retained as his "most precious treasure."[7]

Whipple remarked sourly that the only thing Möllhausen had painted accurately was a Navajo blanket. But most of his sketches for Ives are recognizable, if not strictly representational. Like Ives, Möllhausen succumbed to Romantic reverie, imagined in the chasm's "peculiar formations," for example, "the well preserved ruins of an Indian city," and heavily salted his formal compositions with the conventions of European art and German *Naturphilosophie*. The odd mix of the precise detail and the grand impression so characteristic of the Humboldtean enthusiast suffused the Möllhausen oeuvre. His *Dead Mountain, Mojave Valley* includes a snow-summited peak that exists nowhere along the flanks of the Colorado's floodplain but echoes Humboldt's iconic Chimborazo. That the *Report*'s engraver exaggerated some features recorded on Möllhausen's sketches hardly alters the outcome. What Ives wrote, Möllhausen drew. They even agreed in their prophecies. "With awe I imagined a picture of the rocky canyon of the 'Colorado of the West,' which perhaps for coming centuries will remain a secret to mankind."[8]

But he would make no secret of his western adventures. Returned to Germany, Möllhausen began to shed his acquired roles as naturalist and artist and inaugurated a remarkable career as a writer of romances, a pioneer in the genre of the western. His *Das Halbindianer* [*The Half-Breed*] appeared the same year as the Ives *Report*. His future lay with his wildly popular prose. He settled comfortably at Potsdam, painting in his leisure but writing romances as a career. Along with Karl Friederich May he helped stamp the American West into German Romanticism. Much as James Fenimore Cooper had become the "American Scot," so Möllhausen became the "German Cooper." And while perhaps a minor figure in the American assimilation of the West—sharing the fate of the Ives enterprise overall, its impact shattered by the Civil War—Möllhausen did much to shape European conceptions. America's West, he proved, was also Europe's.

The visual revolution passed to his comrade in arts, F. W. von Egloffstein. Another veteran of the Pacific Railroad surveys, Baron von Egloffstein was more willing to experiment, saw keenly the need for new modes of expression to convey the region's bizarre land sculpture, and carried both poles of the Humboldtean vision to extremes. The schizophrenia between Ives the hydrographic engineer and Ives the literary poseur widened. While the accuracy of Egloffstein's maps increased, the representational quality of his drawings dissipated into Gothic gloom.

A master draftsman, later credited with inventing halftones, Egloffstein had himself engraved Frémont's map of 1853, Beckwith's of 1854, and much of Warren's master map of 1857. Egloffstein the cartographer confronted the radically molded landscapes of the Southwest by inventing clever techniques for depicting the shape of landforms on a map, a system of molding land features by shading that was a useful anticipation of contour mapping.

Yet when Egloffstein the artist examined the same landscape, he gave way almost entirely to the conventions of Gothic illustration. The Canyon became grotesquely exaggerated, a manifesto of Romanticism. In *Big Cañon at the Foot of the Diamond River*, the human figures appear like insects amid gargantuan masses rising to all sides, suggesting, as William Goetzmann has observed, the imaginative engravings of Gustave Doré. Ives alluded to the Canyon as a gateway to the underworld; Egloffstein reinforced that allusion by visually recalling Doré's illustrations to Dante's *Inferno*. Panoramas from the rim show an almost fabulous lack of correlation to any tangible features. Subsequent sketches lose themselves not in a geographic tangle of knotted lines but in a vague dreaminess, almost hallucinatory. A later picture, *Big Cañon*, hedges into phenomenological fantasy. Any objective references, even to diminutive figures, are gone.[9]

It is easy to lampoon Egloffstein's exaggeration, and perhaps he deserves it. But like other Germans of his generation awash in the aspirations of *Naturphilosophie*, Egloffstein could paint impressions with wild hyperbole while simultaneously recording facts in minute, faithful detail. On the Ives Expedition those tendencies finally fissioned, the cartographic science proceeding one way, the landscape art another, with hardly a vestige of their common point of departure.

Yet his art could tap a richer range of associations than the Tower of Seville suggested by the Cárdenas party. There was the literary tradition of a descent to a netherworld, the topographic illustrations and personal narratives of the Second Age, the figurative allusions of Doré, a sanctioned landscape art, especially a landscape art for the Alps. Romantic art and aesthetics had come to emphasize the ascent to light, symbolized by Alpine scenery. The mountain summit was a feature of transcendence, its snow and mist an index of the soul's radiance.

But the Canyon reversed this sense, and Egloffstein struggled to invert not only the aesthetics but the techniques. Alpine conventions forced the eye away from canyon gloom to mountain glory. There is no sublime descent to darkness, no way to celebrate the abyss except as a negation. The thrust is to the summit, and the sooner an exaggerated gorge can carry the eye upward, the better. Oddly, the painting thus looks away from its real subject. The gorge is sealed. On the rim not even this act of inversion was possible; the foreground simply gives way to empty space. There is nothing—no icons, no conventions, no figures—with which to fill it.

Perhaps one can say that the Canyon overwhelmed artistic language. More probably, without input from any contrasting sources of information or the necessity of empirical representation, conventions of art and philosophies of nature overwhelmed the Canyon. Each genre, even in the hands of a common practitioner, went its separate way, splitting as did the Ives Expedition itself. Egloffstein, like Ives, like the expedition, could cope with the river more easily than with the rim. They could share a tradition of explorers plunging through geographic portals into Dark Depths Beyond. They could see the gorge as an inverted summit. There was no equivalent for the rim.

Big Cañon at the Foot of the Diamond River, for all its troubled hyperbole and aesthetic evasiveness, nevertheless remains the first great picture of the Grand Canyon. (Möllhausen's watercolors from Diamond Creek were not published with the report.) It shows what the expedition expected to find, shows the spirit that for better or worse first brought intellectuals to the Canyon, shows how they in turn brought the Canyon into the art of Western civilization. In the end the distortions are less surprising perhaps than the fact that they found any value to the scene at all.

There was no such uneasy compromise for the expedition's chief scientist, John Strong Newberry. The fantastic erosion that had stalled travel and bewildered Alpine art was, for a geologist, a vast commentary on the Book of Nature. Newberry appreciated what the canyons of the Colorado meant, and it was his genius to reconcile their lessons with the larger questions of the age. The Big Cañon moved from freak to exemplar.

Newberry had taken a common path from physician to professor of natural history. Educated at Case Western College, then at Cleveland Medical School, with further studies in France, he field-tested his training with the Williamson-Abbot Expedition in the Northwest, part of the Pacific Railroad surveys. When the Ives Expedition called for his services, he had already established ties with the Smithsonian Institution and George Washington University, part of what was to evolve into a federal science establishment. When his tour with Ives ended at Fort Defiance, Newberry continued on to Kansas. A year later he joined the Macomb Expedition from Santa Fe to the junction of the Green and Grand (Colorado) rivers, elaborating the stratigraphy and erosional mechanics he had witnessed in the western Canyon. During the Civil War he served with the influential U.S. Sanitary Commission. But all this was only a prelude for a man who became one of the patriarchs of American geology. From personal knowledge Newberry could construct a gigantic geologic cross section through North America, and with his expertise Newberry went on to a distinguished career at the Columbia School of Mines, to the directorship of the Geological Survey of Ohio, and to an important role behind the scenes in the politics of national science.

It was Newberry who, more than anyone else, demonstrated that science—historical geology, in particular—was the means by which to incorporate what he referred to as the "Great Cañon of the Colorado" within the canons of intellectual culture. His contribution to mapping Canyon geology is inestimable. He produced the first geologic column of Canyon strata and one of the most important ever drawn, that through which he descended to Diamond Creek. He was the first to intuit the vastness of erosion in the province, an erosion of such magnitude that the Grand Canyon itself was only a minor epilogue. Between his tours with Ives and Macomb, Newberry knew more about the geology of the

Colorado Plateau (which he named) than anyone in the country. He passed this wisdom, along with his maps, to his protégé Grove Karl Gilbert, probably the greatest of American geologists, and from Gilbert, who went back to the Canyon with the Army's Wheeler Expedition in 1871, they became amalgamated with original and borrowed ideas that were to make up the contributions of the Powell Survey during the 1870s and those of its successor, the U.S. Geological Survey after 1879.

"Dr. Newberry" knew the value of what he witnessed, knew that European conventions, even of science, were inadequate to describe it. Boldly he proclaimed that "though valueless to the agriculturalist, dreaded and shunned by the emigrant, the miner, and even the adventurous trapper, the Colorado Plateau is to the geologist a paradise." To one who believed that geology was founded on stratigraphy, the "most splendid exposure of stratified rocks that there is in the world"—as Newberry wrote of the Canyon—was far from worthless, and his account of what he saw there far transcended the picturesque impressions of a traveler's journal or artist's sketchbook. The Big Cañon was a revelation.[10]

His scientific insights had two parts. One concerned fluvial erosion and was specific to geology, although it quickly acquired nationalist overtones. Newberry concluded unequivocally that the plateau's magnificent land sculpturings, both its gorges and its residual monadnocks, *belong to a vast system of erosion, and are wholly due to the action of water. Probably nowhere in the world has the action of this agent produced results so surprising as regards their magnitude and their peculiar character.*" In contrast with reigning thought, on the authority of Lyell, that erosional terraces were the product of marine activity or, following Hopkins, that they were the result of structural catastrophes, Newberry supported and enriched a tradition of fluvialism that argued for the power of rivers as geologic agents, as something more than flumes to move water and debris. Rivers had shaped the land, not merely the landscape its rivers.[11]

There was a nationalist bias to this declaration, for Americans had proclaimed the fluvialist argument most strenuously. James Dwight Dana, a professorial counterpart at Yale and a fellow traveler of the Second Age thanks to the Wilkes Expedition, had argued for the power of fluvial erosion from his observations on South Seas islands. A nation that had several centuries of experience with Niagara Falls needed little urging that rivers could be powerful agents of erosion, but the heartland of geology, Europe, did. That encouragement Newberry supplied magnificently by his contribution to the Ives *Report*, which, significantly, appeared the same year as the monumental tome by Henry Abbot and A. A. Humphreys for the Army Corps of Engineers, *The Physics and Hydraulics of the Mississippi River*. Rivers mattered as much to science as they did to commerce. Almost immediately, thanks to Newberry, the Colorado River became one of the scientifically great rivers of the world, and its canyon not merely an indescribable and impassable tangle of gorges but a textbook case of American fluvialism. The river—not only a Colorado gorged with runoff and spring flood, but that deeper, inexorable flow of geologic time—had made the rim possible.

The question of fluvialism, however, was only a subtext to the larger debate about the antiquity of the earth. If rivers were competent agents of erosion, they succeeded because time was on their side. Conversely, a dramatic demonstration that fluvial erosion had proceeded on an immense scale was an argument for a very old earth. This was Newberry's second achievement. It was no accident that Darwin published the *Origin of Species* the same year Newberry journeyed to the junction of the Green and the Grand, where the great canyons of the Colorado River proper commence. By the time Newberry published for the Ives *Report* two years later, that vision of earth history was not lost on the professor or his readers. Subsequently Newberry assumed a leading role in the reconciliation of Darwinian evolution with theology. In a presidential address to the American Association for the Advancement of Science in 1869, he outlined the terms of compromise, sketching a new version of the argument from Design, this one based on historicism and the belief that "fossils . . . are labels written by the Creator on all the fossiliferous rocks." That same year John Wesley Powell made the first descent of the Colorado's canyons.[12]

The age of the earth mattered to this society, for the Second Age had forced an intellectual crisis. The torrent of discovered organisms, rocks, ruins, peoples, and places demanded an order that geography alone could not impose. Instead the culture turned to history and the belief that time itself obeyed grand patterns. The idea of progress, a ubiquitous organic metaphor that relied on growth, not merely expansion, the sheer immensity of revealed time, all converged on the belief that causality was temporal, the design of the world historical, and the surest explanation one that arranged itself as a successive unfolding of events, stage by providential stage. Increasingly historicism became the common soil of cultivated discourse.

No field of inquiry escaped untouched. In biology, historicism came in the form of theories of development: of the individual, as a pattern of growth and aging; of the species, as evolution; and of the two in syncopation, as proposed by Ernst Haeckel's biogenetic law in which ontogeny reputedly recapitulated phylogeny. It appeared in philosophy, through the dialectic of Hegel; in physical science, with the second law of thermodynamics; and in positivistic social science, by a belief in the laws of progress. Time's arrow, to use Arthur Eddington's striking phrase, was apparently as evident in the physical universe as in the biological and as inevitable in the social world as in the natural. But behind those metaphors and models lay the hard rock of geology, the ground truth of earth time.

The earth sciences shared in this crisis of thought, in fact helped precipitate it and, in the end, helped resolve it. The greatest *terra nova* of the Second Age was geological time—as vast as astronomical space, patient, inexorable, sublime. Like the subtle ether of classical physics, geologic time saturated the natural world from the tiniest crystal to the span of nebulae. Between the late eighteenth century and the mid-twentieth, the known age of the earth increased a millionfold, from less than 6,000 years to more than 4.6 billion. The determination of the exact scale of geologic time and how to organize its unfathomable domain remained the particular province of geology. Upon its conclusions rested the mechanics of organic evolution, and upon those mechanics depended the program of social progress. The age of the earth decided whether Darwinian evolution by natural selection, with its immense drafts of time, was possible. The debate over models of organic evolution informed discussions over moral progress and the future of civilizations.

What the discovery of geologic time had disoriented, the invention of geology helped reorganize. It necessarily did so by appeal to historical methods—created, in fact, methods that became models for other fields of inquiry. The critical breakthrough involved the invention of two types of conceptual chronometer. Both provided a means by which to measure, define, and organize the otherwise unbounded landscapes of time. Both lashed geologic concepts to major scientific themes, a larger universe of philosophical and social discourse. One, the entropy clock, joined geology to the physical sciences through the emerging laws of thermodynamics; the other, the evolutionary fossil, connected geology to the life cycle concepts of the biological sciences, a metaphor that saturated prevailing thought like indigo dye.

The entropy clock treated the earth as a closed physical system. Geologically no new energies would enter, and none would exit. Like the turn of a steam engine, or a bowl of soup placed in a box, energy would flow one way, from a hot core to a cool perimeter. The process was irreversible; each event drove the whole system in one direction; and in the end the earth, or one by one its separate parts, would die an inevitable heat death and survive as a cold, inert slab of stone. Useful work would cease. Energy would exist as entropy. But what gave this melancholy specter promise was that it was possible to measure the flow of loss, which is what an entropy clock could do, and to organize its declination, as natural philosophers and geologists promptly did.

The earth was in fact full of entropy clocks. The progressive cooling ("secular refrigeration") of an originally molten earth, tidal retardation resulting from the friction of the oceans, the decay of solar radiation, the loss of thermal reserves in the earth because of volcanic activity, and so on—all were examples. The most fundamental, however, was the thermal contraction of the planet; to this phenomenon, physical geology could relate all its important topics, especially the evolution of mountains and basins, continents and oceans. For measurements, the infinitesimal calculus was an ideal tool. Like geologic time, it would be unspeakably tiny at any instant of differentiation and suitably vast when integrated over longer intervals. The ultimate entropy clock arrived with the discovery of radioactive decay because this provided both a universal process and an absolute chronology.

The second set of theoretical timepieces relied on another ruling concept of the nineteenth century, evolution. The timeless links on the Great Chain of Being became fossil steps on an evolutionary ladder. Historicizing the Great Chain in this way made paleontology into a reliable and scientifically respectable chronometer. Paleontology became the mathematics of geology, and stratigraphy, the first specifically geologic subject to be organized by fossils, became its mechanics. Just as physics had proceeded by discovering new areas of nature to mechanize, so geology advanced by revealing new strata of the earth, however remote, that it could assimilate into historical chronologies and by discovering new varieties of fossil equivalents in the earth that it could order on the same principles as it had stratigraphy. The earth abounded in preserved relics of its past. By their shape, for example, landforms were believed to reveal a particular stage of development; their forms rendered a kind of geomorphic fossil. So also planetary motions expressed "dynamic vestiges," fossils of the earth's astrophysical origin. Its orbital path, its rotation, its precessional wobble—all testified to the earth's historic evolution as surely as if they were brachipods from a Permian sea or the feathered fossils of *Archeopteryx*.

The various chronometers were themselves similarly scaled and synchronized. Just as the thermodynamic cycle of geophysics could be manifest on a macro and micro scale, so also the evolutionary cycle of biology had its macro and micro versions, from the life cycle of an individual to that of a species—or indeed to the evolution of life itself. Further, the biological and physical conceptions of time could combine. The contraction of the earth by secular refrigeration made for progressively more complex geologic environments; these in turn provided more ecological niches, a prerequisite for progressively more complicated biological evolution; and this additionally made possible the vista of cultural, spiritual, and moral evolution by humans and their institutions. It was simple to project from an evolutionary past to an evolutionary future.

Few of the landscapes penetrated by the Second Age offered so much to geologists so quickly as did the American West. Agricultural wastelands could be scientific and scenic wonderlands. Even beyond enriching its data base, the western experience allowed the earth sciences to intersect the Romantic syndrome of the age. Geology discovered lost earth empires, like ancient mountains worn to oblivion by desert sands in the Triassic or overgrown by fossil jungles of the Carboniferous It exhumed vanished creatures from the Pleistocene and Cretaceous, the forgotten civilizations of natural history buried amid the melancholy rubble of time. It could turn tales of scientific discovery into grand sagas of frontier adventure. It gave the New World natural wonders to compete with decaying castles and lofty cathedrals so fundamental to the cultural adornment of the Old World.

This shared vision of natural and cultural history did not appear instantaneously but emerged piecemeal, as inherited ideas worked through new information and as the new information reworked those ideas. *Geology* did not become a name until the

1780s, and it did not resolve the informing question of the earth's age until the 1950s. But between the time Ives mustered out his troop and worked his reports into print, Darwin published, and Western civilization accelerated a great debate about the character of the natural world and humanity's place within it.

By the early twentieth century so rapidly had geology advanced in the West that an American school of earth science had become the premier of American sciences, its accomplishment recognized through the world. By organizing the realm of geologic time according to certain kinds of historical principles, the American school offered a comprehensive vision of the earth from its microcosm, the mineral, to its macrocosm, the solar system. By focusing especially on landforms, American geology fused philosophy with the practical questions of settlement. Along that sculpted surface earth and humanity met, and with geomorphology the American school made its greatest contributions to science. Nothing, however, epitomized these lessons and this unique experience so fully as did the Colorado Plateau and the illuminated text of the Grand Canyon.

The Ives Expedition journeyed close to the symbolic and geographic center of the Second Age. A complex process finally reached a place commensurate to its ideas and ambitions.

The new explorers to the Canyon saw the scene differently from their predecessors. They had different expectations and the means to satisfy them. In particular, the scientific revolution ensured that the new vision did not merely recycle old perspectives but evolved beyond them. A civilization that pondered the age of the earth would not blink past the Big Cañon, could not dismiss it as strictly the scene for adventurous exploits, or regard it as just another of the West's oversized terrains. The gorge was a trench through geologic time, a place so close to the foundations of earth history that it became axiomatic.

In truth the Canyon was not merely a landscape of time; it was a landscape whose time had come.

INTO THE GREAT UNKNOWN: GRAND CANYON

The political dynamics that made the Ives Expedition possible did not survive the Civil War. Ives joined the Confederacy, Newberry and Egloffstein served the cause of Union, and Möllhausen retired to Potsdam and a shelf of potboiler *Romans*. The Army Corps of Topographic Engineers dissolved in 1863, its members redirected from western rivers and wagon routes to the Little Round Tops and Chickamauga Ridges of the nation's battlefields. Western exploration retired to California, where gold gave geology a reason for state support of the Whitney Survey and where the breadth of a continent intervened between the intrusions of war. The Army never recovered its antebellum preeminence as an exploring institution.

The Second Age, its summit reached, commenced a slow decline. It quickly concluded its most spectacular transcontinental traverses. By 1869 a railroad spanned North America. Between 1867 and 1879 four Great Surveys sprawled across the Far West. The Geographical Survey of the Territories under F. V. Hayden and the Geographical and Geological Survey of the Rocky Mountain Region under John Wesley Powell were civilian, as was the Geological Exploration of the 40th Parallel under Clarence King, though it remained nominally attached to the Army Corps of Engineers; only the later Geographical Surveys West of the 100th Meridian, directed by Lieutenant George Wheeler of the corps attempted to recapture in full the old style of Army reconnaissance. The postwar surveys were often staffed by college men, frequently graduates of the new scientific schools like Lawrence at Harvard and Sheffield at Yale, or of German universities, not by all-purpose naturalists and physicians. By 1879, when the U.S. Geological Survey forced the Great Surveys into consolidation, Henry Stanley had plunged through the Congo, John Forrest and Ernest Giles had crossed Western Australia, and on every continent exploring expeditions of grand reconnaissance surrendered pride of place to routine scientific surveys. Exploration crowded along the surveyed boundaries of Dark Africa's multiplying colonies. For pure discovery only the poles remained.

What did survive, though altered, were the era's intellectual drivers. Darwinian evolution succeeded Humboldtean geography as a model of exploration science. On Newberry's example, history superseded space as an organizing principle, and geology, geography. The age of the earth and the principles by which earth time was organized endured as issues of broad significance. By the year the golden spike was hammered at Promontory Point, Asa Gray and Louis Agassiz had publicly debated both the science and the implied theology of Darwinism, while O. C. Marsh had stepped off a train in Nebraska, uncovered the bones of *Eohippus* from exposed shale, and provided the missing fossil link for the evolution of the horse, a paradigm for organic design. Historical geology based on similar evolutionary principles was not far behind, and that meant that the Grand Canyon, after a ten-year hiatus, would not long stay unvisited.

The lacuna between the Ives Expedition and the Powell and Wheeler surveys was surely not total. Even before Ives a good bit of geographic lore about the canyons along with a dose of rumor and fabrication had circulated among old Southwest hands. At least some trappers knew the region, steamboats had plied the lower Colorado River for at least six years before Ives arrived, and one operator had preceded Ives to Black Canyon by a couple of months. Mormon scouts traveled widely around the region. In 1867 Mormon colonists at Callville, Nevada, rescued a half-dead man on a raft from the river. Whether James White, as he later claimed, had really descended the length of the Canyon or not, he had certainly not intended to if he had, and apart from injecting controversy into the inevitable disputes over priority for the first descent through the Canyon, he contributed nothing to its understanding.

By 1868 the name Grand Canyon had appeared on a railroad survey report by General William Jackson Palmer, Samuel Bowles had confessed that the "great mocking mystery of our geography is the Grand Canyon of the Colorado," and Major John Wesley Powell in outlining his ambitions about descending the river by boat had given as his reason that "the Grand Canyon of the Colorado will give the best geological section on the continent." Certainly something had been added since Ives. The domain of this "mocking mystery" formally began at Lee's Ferry and ended outside Grand Wash Cliffs, with some samplings at Diamond Creek, at Havasu Canyon, and perhaps at a few overlooks on the South Rim. No one had systematically traversed the whole. No one place-name commanded consensus. No dominant image fixed the scene in the public imagination. The eastern Canyon remained terra incognita.[13]

Or it was until Powell penetrated the Canyon twice—once in 1869 and again (to Kanab Creek) in 1871–72. His insight was to float (or bob, row, race, splash, line, and otherwise descend) down the currents instead of fighting them upstream, which demanded steam power. But it was no coincidence that he set his specially modified dories into water at Green River, Wyoming, because here the Union Pacific Railroad intersected the watershed of the Colorado. Between railroad and river, the episodes encapsulated precisely the transition of exploration that would follow.

More than anyone else Powell made the Colorado and its canyons a part of Americana. His personal narrative created the classic expression of the view from the river, the words by which his generation appreciated its revelation, the images by which tourists throughout the twentieth century have understood it. Confronted with the deepening gorges and dark granites of the Grand Canyon proper, Powell announced: "We have an unknown distance yet to run; an unknown river yet to explore. What falls there are, we know not; what rocks beset the channel, we know not; what walls rise over the river, we know not." But ready Powell and his crew were to plunge ahead "down the Great Unknown." These were not the words of literary naturalism or hard science, though they echoed almost exactly Humboldt's pause on his ascent of the Orinoco. "Beyond the Great Cataracts," the young explorer had declared gravely, "an unknown land begins." And in truth Powell's was a final gesture of the Humboldtean adventurer, and his account belongs with Samuel Baker's ascent up the Nile, Henry Bates's travels up the Amazon, and David Livingstone's passage along the Zambezi.[14]

It seemed as if he had uncovered a lost world. Though Newberry had already been to the confluence of the Green and Grand that commenced the Colorado and to the river's debouchment through Grand Wash Cliffs, the stunning bravura of Powell's ambition and his charged, spellbinding prose—filled with commands and heightened by his use of the historical present—made his own voyage of discovery appear unique. A new personality had burst onto the American scene in intimate association with a new landscape. The Major, as he preferred to be called, had deliberately reinforced these impressions by rendering one account out of his two voyages. The *Exploration of the Colorado River of the West*, his official account of the voyages, was finally published under the auspices of the Smithsonian Institution in 1875 after earlier versions had appeared in *Scribner's*.

In terms of securing publicity for a further reconnaissance of the region, the technique was shrewd. A Civil War veteran who had lost his right arm at Shiloh, the Major returned to civilization with the acclaim normally reserved for war heroes or, in a later age, for Mercury astronauts. The Geographical and Geological Survey of the Rocky Mountain Region, based out of Kanab, Utah, became a reality, cobbled together from congressional funds, War Department rations (courtesy of U. S. Grant), and Smithsonian oversight. By the summer of 1872 Powell and others were traveling to the Canyon from Kanab, the first contact for the North Rim. Survey parties explored the High Plateaus, discovered the Henry Mountains, worked out the hydrographic history of the Great Salt Lake, and laid an empirical foundation for later political reform of the public lands.

Publicity and politics were mutually reinforcing. The geology was an open book. Erosion exposed geologic structures as cleanly as dinosaur bones. Strata spread out like illuminated parchments. What Newberry had suggested, Powell confirmed. But Powell the politician recognized that art could reach a larger audience than science and arranged for Thomas Moran, fresh from his triumphs with the Hayden Survey at Yellowstone, to join him. Riding popular acclaim as he did the Colorado rapids, Powell acquired a prominence sufficient to eventually become director of both the Bureau of American Ethnology and the U.S. Geological Survey.

Yet in many respects Powell's celebrated account, like Powell himself, was an anachronism. Powell the explorer was a throwback to the Humboldtean tradition of the personal narrative and the explorer as Romantic hero, much as Powell the politician was a throwback to the agrarian philosophy of the Jacksonian era and its reformist enthusiasms. His report was a journal, though one helped mightily by the character of the river on whose powerful current the narrative could travel. It segregated its science from that story as much as Ives had. Powell's party had consisted largely of adventurers—Civil War veterans, displaced mountaineers, relatives—and they were led by a self-taught naturalist, though he made himself into one of the great scientific amateurs of the nineteenth century. The original Powell Survey and its immediate successors were a far cry from the sophisticated corps fielded by Clarence King or the armada of naturalists brought

west by Ferdinand Hayden or even the cartographic expertise ex-
emplified by Lieutenant Wheeler and his Army engineers. Powell's literary characterization of the Canyon was largely of a piece
with that of Ives, or Pattie, or even Garcés. The gorge was confining; Powell referred to it as "our granite prison."[15]

In reality there was little science in the *Exploration*, and a good
deal of stunt, and of melodrama, and of rhetoric. The science
came later, after Wheeler had brought his party up by river to
Diamond Creek and after G. K. Gilbert, the crown jewel of
Wheeler's scientific contingent, had begun a close, lifelong exchange with Powell. The science appeared in the 1875 edition of
the *Exploration* and, even more, in the companion *Geology of the
Uinta Mountains* published a year later—the same year, in fact, as
Newberry's long-withheld report for the Macomb Expedition.
Curiously, the Major's insights into the structural geology of the
region came from the dramatic intersection of the Uintas with the
Green River, like a melon cut in half, not from the complex passage of the Colorado through the Kaibab.

The process was not dissimilar, however, and its description
was Powell's great service to geology. In both cases the river had
plunged through an upward-swelling earth, cutting a deep canyon, a process that Powell labeled "antecedence" because the
river was antecedent to the mountain and a process that with his
gift for homely analogies, he likened to the action of a log (the
mountain) being raised into a buzz saw (the river). It was a position entirely in keeping with American fluvialism and one that a
later Powell associate, Clarence Dutton, formalized into the law of
the persistence of rivers. It was the river that excited Powell's
imagination, and it was rivers—or more properly their water—
that animated his subsequent career as a geologist and administrator of the Geological Survey.

The law of the persistence of rivers. The phrase might as well
apply to Powell's *Exploration*, for it not only permanently forged
the perspective of the Grand Canyon from the river but confirmed that the river was the Canyon's essence. The flow of one

provided the narrative structure for the other. In fact, Powell's account speaks little of the Canyon and much about the prisonlike defile down which his group labored. The Grand Canyon was noteworthy for having the worst rapids, not the best scenery. Powell said as much. "All around me are interesting geological records. The book is open, and I can read as I run. All about me are grand views, for the clouds are playing again in the gorges. But somehow I think of the nine days' rations, and the bad river, and the lesson of the rocks and the glory of the scene is but half seen."[16]

Only much later, after he had resigned under pressure from the directorship of the Geological Survey, after he had helped conduct an International Geological Congress to the South Rim, after he had undertaken to expand and revise the *Exploration*, did Powell climb out of that confining gorge and survey the rim and declare the scene "the most sublime spectacle on the earth." By then he had the advantage of others' aesthetics, as he did their science. "A year," he announced, "scarcely suffices to see it all." But "if strength and courage are sufficient for the task, by a year's toil a concept of sublimity can be obtained never again to be equaled on the hither side of Paradise."[17]

John Wesley Powell rode the rapids of the Colorado to prominence. His voyage became as much a saga of American discovery as Lewis and Clark ascending the upper Missouri or Jedediah Strong Smith circumnavigating the Trans-Mississippi West. His vision remained fixed to the river and through it to the power of fluvial erosion as a geologic force and, more broadly, to water as a social force in western settlement.

In Canyon historiography, Powell has remained the prime mover. What the river gave, he returned. Once invested with power, Powell ensured that the Grand Canyon would share his prominence as a kind of monument to America's westward destiny. Without him, it is unlikely that the Army would have sent

G. M. Wheeler to recapitulate the Ives Expedition, that Thomas Moran and William Holmes would have drawn and painted their Canyon panoramas, that Clarence Dutton and Grove Karl Gilbert would have amplified the legacy of Newberry and interrogated the geomorphology of the Colorado Plateau, that the Grand Canyon would have become preeminent among American landscapes.

Without Homer the Trojan War would have disappeared amid the endless, unrecorded conflicts of ancient history, and the sack of Troy would be indistinguishable among the thousand-year strata of rubble and ash that layer its site. Without Homer there would have been no Achilles, Agamemnon, Helen, or Odysseus; no *Aeneid* to transfer the epic elsewhere; no Heinrich Schliemann to search out and exhume its artifacts. So it was with John Wesley Powell and the Canyon: the Colorado River and its gorges had found their poet laureate, and an American bard, the saga he would sing for his career.[18]

AGAINST THE CURRENTS: RETURN TO BIG CAÑON

By the time Powell readied his second voyage down the Colorado rapids, the Army Corps of Engineers was organizing for a second trip up. Lieutenant George M. Wheeler raised again the banner of the Army explorer. After two seasons in the field with modest parties, the Geographical Survey West of the 100th Meridian expanded in personnel and ambitions. In 1871 it logged more than six thousand miles through the Great Basin and the Southwest in a furious effort to reclaim Army preeminence. Not surprisingly that revival began where Ives left off. This time the Army engineers transported their boats through the gorges from Black Canyon to Diamond Creek.

Ultimately the Wheeler Expedition succeeded in sailing, poling, and dragging three boats from Camp Mojave to Diamond

Creek, where they rendezvoused with the memory of Ives and departed. Wheeler asserted that he had finally answered the riddle of the Colorado River's limit of navigation—not that anyone doubted that it ended at the great canyons of the plateau. And then, echoing Ives's misjudgments, he studiously slighted Powell and pontificated that "the exploration of the Colorado River may now be considered complete." That proclamation was as flawed as Ives's.[19]

Yet the expedition had some impressive achievements. For a photographer Wheeler had the services of Timothy O'Sullivan, a Mathew Brady protégé and a veteran of the King Survey, recently returned from the Navy's Darien Survey Expedition to Panama. Wrestling heavy cameras and glass negatives, commandeering the services of a whole boat, O'Sullivan produced the first photographs of the Canyon. For geologists the expedition had men like Archibald Marvine and G. K. Gilbert. For artists, regrettably, the Wheeler Survey had none, and for its narrator, even more regrettably, it had only Wheeler himself. Wheeler's clumsy prose, crimped vision, and political posturing made even his survey's best work look flaccid and sometimes fatuous. Its accomplishments failed to match its promise.

The survey's real coup was Gilbert, a figure destined for geologic glory. The 1871 expedition was his introduction to western exploration. With two years' apprenticeship under Newberry behind him—both in Ohio fieldwork and as a kind of aide-de-camp for the professor's winters at Columbia—Gilbert came with his old boss's insights in his head and his maps in his hands. To Gilbert belongs the formal naming of the Colorado Plateau and the Basin-Range Province and the first systematic delineation of their unique features. He added other names to Newberry's stratigraphic column—the Redwall limestone, for instance. He worked out the basic mechanics behind the structural geology and volcanics of the region. In 1872 he surveyed the region from out of

Kanab, visiting the North Rim and descending Kanab Creek to the river. Later in Salt Lake City, he met Powell, and they began their great collaboration—"swapping lies," as Gilbert put it. In 1875, irritated by Army methods of reconnaissance and mapping that subordinated geology to topography, Gilbert transferred to the Powell Survey, where, ironically, he spent his first field seasons remapping areas that Powell's amataeurs had bungled.[20]

Powell reaped what Wheeler had fecklessly sown. Gilbert's thoughts on the structural geology of the plateaus, on rivers, on rainfall and lake levels in the Great Basin—all programs that had begun under the Wheeler Survey—were released under Powell's rubric and to Powell's fame. Gilbert's final report for the Wheeler Survey was published in 1875, though its ideas had been expressed in progress reports as early as 1872. For Powell he summarized important ideas on the structural geology of the plateaus and the mechanics of fluvial erosion in the classic *Report on the Geology of the Henry Mountains* (1877), and he supplied most of the scientific research behind Powell's manifesto for land reform, *Report on the Lands of the Arid Region of the United States* (1878). Gilbert continued many of the Powell Survey themes as director of the Division of the Great Basin after the U.S. Geological Survey was established. Under Powell's regime at the Geological Survey (1881–94), he faithfully surrendered original research for administrative chores, eventually becoming chief geologist. Only with Powell's resignation from the survey did Gilbert return to the field, and curiously only after Powell's death in 1902 did he revive the brilliance of his early years. The executor to Powell's will and his boss's first memoirist, Gilbert apparently found it necessary to lay the patriarch to rest before he could return to his own special studies.

In his science too Gilbert organized, rationalized, cleaned up after the charismatic and often careless Powell. Though Powell dramatized the concept of antecedence, it was Gilbert who systematically translated fluvial erosion into Newtonian mechanics. Where Powell turned the river's gorge into a purple-prosed prison, Gilbert defined it as a gigantic flume engaged in the business of moving debris. Where Powell likened river and Uinta Mountains to a buzz saw and log, at the Henry Mountains Gilbert compared the laccolith (a mountain type he named and first described) to a hydraulic piston and applied some elementary mathematics to the forces involved. It was Gilbert, not Powell, who consolidated the American fluvialist position, Gilbert who translated river silt and sand into equations, Gilbert who wrote a Euclidean treatise on erosion and structure that served as the *Elements* of the American school. Though he had to wait more than thirty-five years after the Wheeler Expedition to do it, and though they would be based on mining debris in the Sacramento rather than silt in the Colorado, Gilbert's flume experiments on the University of California campus marked the beginning of experimental sedimentology and the scientific investigation of debris transport by streams. Conceptually the Colorado River Gilbert entered at Fort Mojave ultimately debouched over the tidal bar outside San Francisco's Golden Gate.

Its experience with Gilbert was in many ways typical of the Wheeler Survey at large. Time and again, the lieutenant found himself outranked by the Major. Devoted to the Army and the tradition of the military explorer, egged on by the combative chief of the Army engineers, A. A. Humphreys himself, Wheeler searched eagerly for a spectacle that would put his survey in the public eye. Powell had the Colorado canyons, Hayden the Yellowstone, King the Sierra Nevada. With a mixture of boldness and temerity, Wheeler sent field parties to all those areas—too late to claim priority, yet early enough to be charged with duplication.

His grand gestures always fell short. When he labored up a mountain in the Sierra Ancha and satisfied himself that no white man had ever been there before, no white man seemed to care either. A July traverse across Death Valley ended in sunstroke and ridicule. His expedition up the Colorado River in 1871, clearly intended to challenge Powell's presumption and to recapture the glory days of the corps, instead left him fighting against the currents of popular opinion. Wheeler was traveling the wrong way. His Canyon experience is surely the least remembered (or valued) of any of the era.

Though more solid work came out of his survey than out of Powell's, that fact hardly mattered. Powell was a master of prose and politics who brought his work to the public; Wheeler's tomes struggled to make it from government files to library shelves. No one mapped more of the West than Wheeler, and with the possible exception of the King Survey, no one mapped it with more skill. But his turgid prose and prosaic straining for effect looked florid, even silly, next to Powell's taut narratives and dramatic simplifications. Of the Colorado canyons Wheeler declaimed in 1889, "They stand without a known rival upon the face of the globe, and must always remain one of the wonders, and will, as circumstances of transportation permit, attract the denizens of all quarters of the world who in their travels delight to gaze upon the intricacies of nature." They would indeed come to the Canyon, but not on the strength of Wheeler's numbing prose.[21]

After 1879, broken in health and granted a disability discharge from the Army, embittered over rivalries with "scientific filibusters" like Powell, his beloved survey abolished in favor of the civilian U.S. Geological Survey, Wheeler retired into anonymity, laboring for another decade over his accounts. By the time his *Geographical Report* appeared in 1889, few people cared about the survey or even remembered it. His work in the Canyon was completely overshadowed by that of Powell; even Gilbert's monograph *Geology* is the least well known of all Gilbert's writings. Instead Gilbert's ideas—and those of Wheeler's 1871 expedition—were absorbed into the corpus of the Powell Survey and the works of John Wesley Powell himself.

* * *

Surely this latter exchange was not unilateral. In eulogizing Powell, Gilbert declared that Powell's greatest contributions were those he never published but passed out freely as ideas to his colleagues. Clarence Dutton said much the same: among himself, Gilbert, and Powell there was such a full and frank circulation of ideas that no one knew who had originated what concept. Yet the self-effacing Gilbert, the real successor to Newberry in the plateau province, should not be forgotten. He lacked the pulpit oratory of Powell that would make the *Exploration* a perennial classic or the suppleness of mind and grace of prose that did the same for Dutton's *Tertiary History of the Grand Cañon District*. But the science of erosion in the Colorado Plateau was spectacularly his.

Bailey Willis, who watched both men closely on the national survey, spoke of Gilbert as "Powell's better half. Perhaps no one else ever thought of them in that way, but in constant relations with the two I learned to know how much Gilbert, the true scientist, contributed to the geological thinking of Powell, the man of action. I do not think that they themselves were conscious of the degree to which the latter absorbed and gave out as his own ideas that the former had silently passed through." Where Powell preferred a dramatic revelation, Gilbert proceeded by systematic contrasts between known and unknown, exploring analogies that linked new with old, that bonded a novel science like geology to traditional physics. Gilbert, for example, ignored Powell's intricate taxonomy of structures and streams because they were not "founded on principles of causation, and cannot therefore be assumed to be final." Instead he searched for those informing principles, finding them by analogy to mechanics rather than, as Powell had, by analogy to paleontology and stratigraphy.[22]

So too, while Powell brought the Canyon into exploration lore, Gilbert transported it into natural science. The American school of geology boosted by William Morris Davis celebrated fluvialism, mountain building, and geomorphology, all dramatically revealed in the plateau province and published by that fabled triumvirate

Powell, Dutton, and Gilbert. Yet Powell's bravura had behind it Gilbert's scientific muscle, and Dutton's word paintings of Canyon land sculpture decorated the nearly mathematical perspective of Gilbertian geomorphology. Gilbert went on to a celebrated career: a pioneer of experimental geology and the philosophy of method, the discoverer of two mountain-building processes and enunciator of the modern theory of the moon's origin, the third American to receive the coveted Wollaston Medal from the London Geological Society, a member of the National Academy of Sciences, and the president of numerous scientific societies, including the Geological Society of America, the only man twice so honored. His *Lake Bonneville* was awarded pride of place in the U.S. Geological Survey's monograph series, much as his career burned inextinguishably in the institutional memory of the survey. For its centennial the survey established a G. K. Gilbert fellowship to honor its outstanding practitioners.

Gilbert left to Powell the grand gesture and the purple-prosed popularizing of the Canyon. Within Powell's public career, he buried his own. No memorial to him rises along the rim. Yet the science he did so much to validate was what, in the critical years, did so much to validate the Canyon. Geology remembered. Over evening campfires in 1965 the geologists of the USGS Colorado River Expedition, the last survey before Glen Canyon Dam closed the free flow of the Colorado through the Grand Canyon, debated the question of who was America's greatest practitioner of what had surely become America's greatest science. They chose Grove Karl Gilbert. The setting as much as the selection was exactly right.[23]

A GREAT INNOVATION: GRAND ENSEMBLE

The Second Great Age of Discovery unveiled—more than that, endowed with meaning—many landscapes. But the Canyon may have been its most heroic creation because it was the strangest of its encountered scenes. Meaning requires a cultural context; none existed for so unprecedented a place. Such a context, moreover, requires continuities. But logic and language strained against features so bizarre and unexpected.

Yet the deed was done. Powell positioned the Canyon squarely in the vogue of the Humboldtean explorer and the context of a Civil War veteran, feverishly pursuing the dark unknowns of the West, breaching plateaus with the élan of an assault against the fortifications at Vicksburg. Gilbert fused the region to natural science. He established its significance through close analogies from known to unknown, solving its geologic puzzles with almost algebraic rigor. The Colorado Plateau became not merely a textbook demonstration but a source of exemplars and first principles. It remained to Clarence Dutton, however, to close the cultural triangle by bonding the Canyon to art. What the graphic artists of the Powell Survey did with pencil, paint, and wet plate, Dutton did with words. The Canyon became a place where the great themes of the age found monumental expression.

More than the others, Dutton appreciated the core problem, the oddity of the place. He understood that the Canyon was a grand ensemble whose effect was greater than the sum of its parts. And he discovered a means by which to celebrate that strangeness and to hold those pieces together. The *Tertiary History of the Grand Cañon District*, which Dutton published in 1882, was itself an extraordinary ensemble of the science, aesthetics, cartography, painting, photography, illustration, and ideas that had animated the intellectual and imperial expansion of America.

Dutton created an aesthetics without a popular substrate, without historical antecedents or cultural transitions, and he did it from the rim, without the narrative flow that a river trip intrinsically imposes. From the rim the perspective is more diffuse; there is no obvious focus because no central figure dominates and the eye sweeps across a vast horizon, almost an inversion of typical perspective; a narrative of the panorama has no intrinsic structure. Instead Dutton discovered in the flow of geologic time, in eras of deposition and erosion, an equivalent to the river's pools and rapids, and in the organization of time that preoccupied so much of the nineteenth century's intellectual elite, he discovered an informing conceit for an epic history of the earth.

He accepted the essential strangeness of the scene. Instead of feeble allusions or forced analogies, a soothing similitude to contemporary tastes, Dutton boldly asserted the peculiarity of the place, its unblinking uniqueness. There was really nothing else like it anywhere. It might have come from another planet or another existence. Had it been sculpted in central Europe, it would have shaped European aesthetics as much as the Alps. But it wasn't, and it didn't, and instead it stood outside virtually every convention of perception and understanding that millennia of Western art and philosophy had nurtured. John Muir, no amateur at cultivating landscapes, observed that "no matter how far you have wandered hitherto, or how many famous gorges and valleys you have seen, this one, the Grand Cañon of the Colorado, will seem as novel to you, as unearthly in the color and grandeur and quantity of its architecture, as if you have found it after death, on some other star. . . ."[24]

Dutton appreciated precisely the problem. "Great innovations, whether in art or literature, in science or in nature, seldom take the world by storm," he explained. "They must be understood before they can be estimated, and must be cultivated before they can be understood." No one has stated the fundamental issue better; probably no one ever will. By the time Dutton finished his inquiry, the Grand Canyon, essentially unknown twenty years before, utterly outside canons of art, literature, and science, had become an inextricable part of all and in fact became itself an exemplar of landscape aesthetics.[25]

Dutton's was the great innovation, the view from the rim. He put ideas behind the imagery. If, in Canyon historiography, Powell is the indispensable man, Dutton is the indispensable mind. For Powell the Grand was one of many canyons, and had it been absent, his river adventure—and the attendant geology and drama—could have continued and carried him to fame on the revelations from the Canyon of Lodore and Glen Canyon and the drama of Separation Rapids. For Dutton, however, Grand Canyon was a climax, without which the other landscape parts had no defining presence, no informing narrative. The *Exploration* took its narrative flow and rhythms effortlessly from the river. The rim offered no equivalent order. It was Dutton's insight to find that requisite flow in the passage of geologic time and to order it by an evolutionary progression of landforms. A Canyon panorama was not a confusion of lithic shapes and an empty sky. It told a story; it had a structure by which the mind could organize the eye; its geomorphic profligacy constituted in fact an aesthetic ensemble.

Wallace Stegner observed that Dutton has become "almost as much the *genius loci* of the Grand Canyon as Muir is of Yosemite." With few exceptions, those who came to the Canyon saw it, as Dutton did, from the rim. Those who followed Powell down the river (like Robert Brewster Stanton), or put their own voyages before the public (as did Frederick Dellenbaugh), or who tried to compete with Powell's narrative (as did Wheeler) wrote too late and, for all their labors, added too little. The river's narrative drive was powerful but singular in its effects, limited to one tale, and for long decades inaccessible in its popular recreation. For nearly a century after Powell's feat, the American public connected to the Canyon through its overlooks, not its rapids. The Santa Fe Railroad brought tourists and pilgrims to the South Rim, not to the Inner Gorge. Accordingly, as Stegner concluded, "although it is Powell's memorial to which the tourists walk after dinner to watch the sunset from the South Rim, it is with Dutton's eyes, as often as not, that they see."[26]

* * *

Who was Dutton, and how did he come to see what others had overlooked? The unorthodoxy of his vision reflected that of his career.

Clarence Edward Dutton was a Yale man (class of 1860); an athlete (he rowed crew); a litterateur, who had won the literary prize as a senior; an enthusiastic scientist, attracted to mathematics and chemistry and through them to geology; a Civil War veteran, brother to a West Point engineer, a volunteer who made the Army his career, despite some "pretty rough service," as he put it; an accomplished conversationalist, who could lecture extemporaneously on the silver question or on South Seas ethnography with equal relish; and an agnostic, who after spending two weeks at the Yale Divinity School, in his words, left before he was thrown out. He had a soldier's bearing and a scientist's mind. He was, in brief, the epitome of that great transition in American exploration between the Army engineer-explorer and the civilian scientist. He was both.[27]

A congenital polymath—he once described himself as "omni-biblical"—Dutton combined concentration with versatility. He loved cerebral games, especially chess, although eventually he abandoned the game because the sleepless nights he spent puzzling over it threatened to undermine his health. Instead he transferred that intellectual passion to other subjects with their own riddles: ordnance, chemistry, and geology. His memory was prodigious. He composed most of his manuscripts in his head, then wrote them down or dictated them in marathon sessions during which he stalked back and forth across a room, alternately speaking and puffing on an omnipresent cigar. In one such session he dictated eleven thousand words, of which two minor corrections were made in later revisions. No doubt this procedure helps account for the exceptional grace and conversational tone of his writings.

During the war he transferred from the Connecticut Volunteers to the Ordnance Corps of the regular Army, only one of three to pass the examinations. When the war ended, he remained as a captain of ordnance assigned to the Watervliet Arsenal at Troy, New York. He soon put his aptitude for mathematics and natural science to use. The nearby Bessemer steelworks stimulated an interest in the chemistry of molten materials, a good transition by which an expert in ordnance could move into volcanoes. Dutton's paper was the first analysis in the United States of the Bessemer process. Meanwhile the Albany Paleontological Museum, which Dutton frequented, encouraged the study of geologic phenomena. His early tutors were James Hall and R. P. Whitfield. Their specialties, however, would not be his, though Dutton did propose a new mechanism for Hall's concept of the geosyncline. But apart from their interest in historical geology, Dutton was too much an educated man of his time to ignore the evolutionary syntheses being proposed for field after field of intellectual endeavor and for which earth history was both paradigm and precondition.

In 1871 he was reassigned to Washington, D.C. He quickly "insinuated" himself into the burgeoning scientific clubs, especially the Philosophical Society, and, as he recalled, "cultivated the Survey men and became well acquainted with Powell." Thanks to Powell's connections with U. S. Grant, the War Department detailed Dutton out to the Powell Survey in 1875. The arrangement lasted for fifteen years. For half of each year, thanks to special acts of Congress, Dutton worked in the field for Powell and later for the U.S. Geological Survey. For the other half, he returned to the Ordnance Corps and Washington society.[28]

Thus began a long furlough, a wonderful intellectual adventure for an inveterate cigar smoker who loved to read Macaulay and Mark Twain and who, throughout it all—amid the most arduous fieldwork and the most sophisticated theorizing—maintained an irrepressible touch of irreverence. From the High Plateaus Dutton wrote Powell that he had solved a long-vexing problem: "how high and steep and rough a hill a mule can roll down without getting killed." But he also moved in high society, and it was Dutton who in 1878 recommended to Powell that they found the organization that became the Cosmos Club.[29]

As his name for the club suggested, Dutton's was a Romantic intelligence. What attracted him to a subject was a sense of its scope, vision, and unity. It was his desire, as he put it, "to discuss only such geological fields as present a series of facts which can be grouped together into a definite, easily comprehensible whole, and to avoid a subject which has, so to speak, neither head nor tail to it." He found that "comprehensible whole" in evolutionary laws.[30]

Through them he could organize his travels and relate his favorite subject, vulcanism (the study of volcanoes), to igneous petrology, earthquakes, isostasy, and the "larger questions" of earth history. From his analysis of volcanic processes, he was able to bring igneous stratigraphy and landform studies into conformity with theories regarding the geophysical evolution of the earth. After Powell's ascension to the directorship of the Geological Survey, Dutton became chief of the Division of Volcanic Geology. As a result of expeditions to Hawaii, Central America, and the major volcanic fields of the Far West, he became America's outstanding vulcanist, a theoretician in the field that underwrote the major laboratory discipline in American geology.

It led to other intellectual fields as well. As a result of his investigation of the Charleston earthquake, he was credited with bringing the "new seismology" to the United States and was memorialized in the first *Bulletin of the Seismological Society of America*. His report to the Senate on the volcano and earthquake potential of Nicaragua was an important document in the decision to build an isthmian canal through Panama instead. Still searching for a "comprehensible whole" to questions of mountains, eruptions, and tremors, he published a celebrated paper in 1889 on the subject of gravitational equilibrium in which he dusted off an old speculation by the astronomers Herschel and Airy, renamed it "isostasy," and used it to criticize the contractional hypothesis, then the primary theory of earth geophysics. For his achievements Dutton was early elected to the National Academy of Sciences. In 1906, addressing the academy, he rushed to apply the new crustal

FIRST IMPRESSIONS

Two visions from the Ives Expedition: At left, F. W. Egloffstein's *Big Cañon at Mouth of Diamond River*, redrawn by J. J. Young, and below, Balduin Möllhausen's *Mouth of Diamond Creek, Colorado River, View from the South.* Since Möllhausen's watercolor was not published with the official report, Egloffstein's *Big Cañon* begame the primary source of visual appreciation.

THE SECOND AGE GOES WEST

The Ives Expedition's celebrated steamboat, *The Explorer,* as recorded by Möll-hausen (above) is a curious blend of American genre painting and German *Naturphilosophie*—with noble savages in the foreground, improbable land-scapes to the background, and the crew rather resembling George Bingham's merry flatboatmen on the Mississippi. *Dead Mountain, Mojave Valley* by Möll-hausen, and redrawn by J. J. Young (opposite, top), shows an improbable feature of the Colorado River region, but one that echoes the iconography of the Humboldtean explorer, as recorded in this F. G. Weitsch painting (opposite, bottom) of Humboldt, every inch the European savant, bringing the light of modern learning to Ecuador with a radiant Mount Chimborazo in the background.

CANYON COLUMNS: SCIENCE AND SENTIMENT

Egloffstein recorded the western Canyon as a Gothic spire (*Big Cañon*, left), while John Strong Newberry recorded the features as geologic strata in one of the most famous stratigraphies from the Grand Reconnaissance (below). Ives himself considered Newberry's report the "most interesting and valuable result of the explorations."

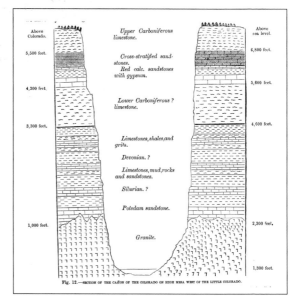

Above Colorado.			Above sea level.
5,500 feet.	Upper Carboniferous limestone.		6,800 feet.
	Cross-stratified sandstones. Red calc. sandstones with gypsum.		5,600 feet.
4,300 feet.	Lower Carboniferous ? limestone.		
3,300 feet.	Limestones, shales, and grits.		4,600 feet.
	Devonian. ?		
	Limestones, mud, rocks and sandstones.		
	Silurian. ?		
	Potsdam sandstone.		
1,000 feet.	Granite.		2,300 feet.
			1,300 feet.

Fig. 12.—SECTION OF THE CAÑON OF THE COLORADO ON HIGH MESA WEST OF THE LITTLE COLORADO.

GEOLOGISTS IN THE GORGE

The critical personalities in the transition from geographic reconnaissance to geological science: John Strong Newberry (above left), John Wesley Powell (above right), and the link between them, Grove Karl Gilbert (below), studying rocks while others sleep.

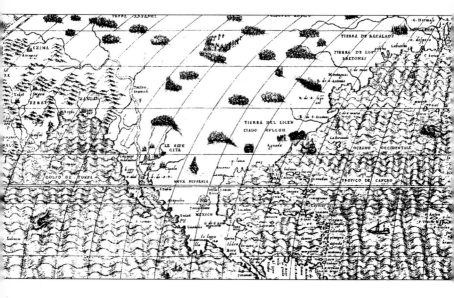

THE CARTOGRAPHIC CANYON I: CAÑON AND QUIVIRA

ABOVE: The Gastaldi map of 1546 (detail), one of the greatest of early New World maps, with the Colorado River dominating the known terrain of North America.

OPPOSITE, TOP: The Miera map of 1777, summarizing the cartography of the Dominguez-Escalante expedition. Miera obscures the main region of the Canyon, which the expedition bypassed, though he refers to Marble Canyon as *mui escarpada* and shows the Little Colorado. Like most maps of the period, its basic information is hydrographic and ethnographic.

OPPOSITE, BOTTOM: Garcés's map of his journey to the Moqui (1776–77). The map accurately portrays the major drainage of the Colorado River, though—because Garcés did not follow the river continuously but leap-frogged from settlement to settlement—it misses the great bend above Black Canyon and the rivers to the north. It does, however, show two streams entering from the south into what Garcés called the Puerto de Bucareli, surely Diamond Creek and Havasu Creek. Garcés named the Puerto after Antonia Maria Bucareli, Viceroy of New Spain.

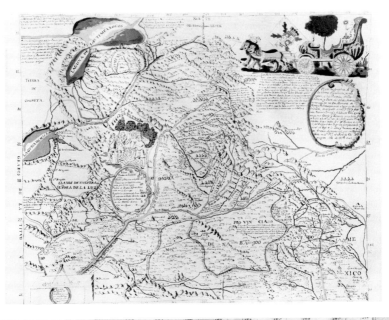

Mapa
Formado sobre el Diario del Viage que hizo el P.º
Fran.ᶜᵒ Garcés al Rio Colorado, S.ᵗ Gabriel y
Moqui en 1777.

Escala de 50 Leguas de a 5000 Varas Cada una.

THE CARTOGRAPHIC CANYON II: GRAND RECONNAISSANCE

LEFT: Humboldt eyes the Canyon—the 1811 map (detail). Humboldt collated Spanish sources into a great map of New Spain. Through it he popularized the cartographic device of hachures as a means of indicating mountains, one of his many innovations. The Puerto de Bucareli is kept, probably on the basis of the Garcés map, but its location has shifted from the Colorado River to the Little Colorado.

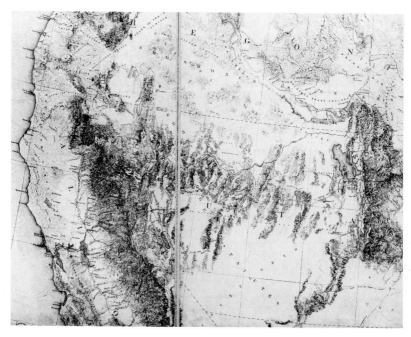

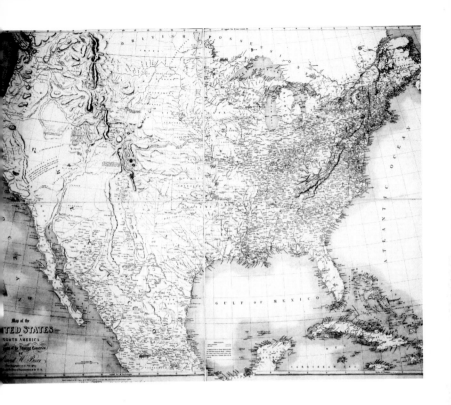

OPPOSITE, BOTTOM: The David H. Burr map of 1839 (detail). Relying on information generated by the fur trade, the map is reasonably complete for the Rocky Mountains and northwest; but, except for Jedediah Smith's trek along the Old Spanish Trail, the southwest is a void. Until official surveys that followed the Mexican War, Humboldt's map remained the primary reference for high culture.

ABOVE: G. K. Warren's map of 1857. Consolidating the Pacific Railroad Surveys of the 1850s with previous information, Warren's map shows with some accuracy the country traversed by the proposed transcontinental railroad routes and of course the Old Spanish Trail. The Colorado River is sketched very lightly; its east-west transit is recognized and the Little Colorado is located correctly, but nothing identifies the Canyon. Still, this sophisticated map shows major advances in scientific cartography. Within its format the context clearly exists for a landscape feature like the Canyon, devoid of ethnographic identity or commerical hydrographic significance.

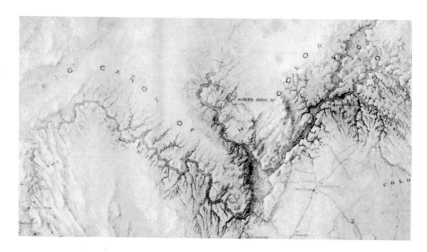

THE CARTOGRAPHIC CANYON III: HEROIC AGE

TOP: Egloffstein's magnificent map for the Ives Expedition, 1861 (detail). The map marvelously captures the serpentine river and its gorges, though the overall hydrography is flawed by confusing tributaries with the main Colorado. The contrast with Egloffstein's landscape art is striking.

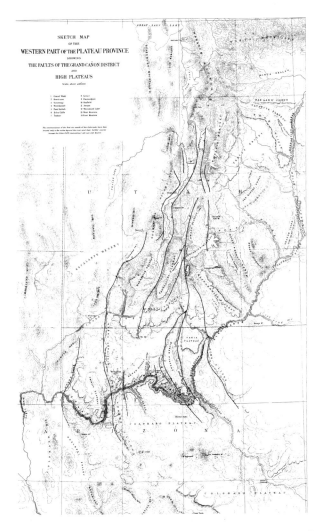

OPPOSITE, BOTTOM: The Wheeler Survey map of 1871, incorporating the 1871 Canyon expedition. Though much is still missing, the larger hydrography of the Canyon is at last accurately represented, thanks especially to the two Powell surveys.

ABOVE: Clarence Dutton's map of the High Plateaus and Grand Canon district, showing clearly the main landscape feature and the geologic structures that define them. Canyon cartography is beginning to incorporate Canyon time.

RIVER AND RAIL:
AMERICA'S CONTINUING MANIFEST DESTINY

Powell's second expedition sets off down the Colorado River (above). The first expedition pushed off the same year that the transcontinental railroad completed its tracks, celebrated in this Currier & Ives print, *Across the Continent, "Westward the Course of Empire Takes Its Way,"* by F. F. Palmer (below). The boats launched at Green River, Wyoming, where the Union Pacific and the Colorado River intersected, geographically as well as symbolically.

THE PHOTOGENIC CANYON

LEFT: Jack Hillers on the High Plateau with complete photographic outfit, a considerable burden.

BELOW: The equipment commanded a full boat on the Wheeler Expedition, as shown by Timothy O'Sullivan's self-portrait of *The Picture*.

SCIENTIFIC SUBLIMITY: THE PANORAMIC CANYON OF WILLIAM HOLMES

ABOVE: *The Temples and Towers of the Virgen*

OPPOSITE, TOP: *The Grand Cañon at the Foot of the Toroweap*

OPPOSITE, BOTTOM: *Panorama from Point Sublime, Looking East*

Superb examples of topographic art, and perhaps the finest landscapes ever made of the Grand Canyon region, east and west.

Captain Clarence E. Dutton   William Henry Holmes

The Grand Ensemble:
A Great Innovation and Its Innovators

Thomas Moran

heat source discovered in radioactive decay to old problems in vulcanism. A lifetime of harmonizing field and theory made Dutton a major figure in the reconciliation of geological and geophysical theories of global evolution.

Yet it is for his landform studies on the Colorado Plateau that Dutton is best remembered. His achievement was twofold. Along with Powell, Gilbert, and Newberry, he brought the strange spires, majestic cliff facades, upthrusting mesas, and fabulous canyons into the realm of scientific explanation. Together they consolidated a coherent geomorphology, what became the basis for the American school of geology. But whereas Newberry announced, Powell dramatized, and Gilbert reduced to Euclidean logic, Dutton sought an artistic appreciation. He cultivated a critical aesthetic meaning for a physical geography that lacked referents and folk attachments. To overcome what he considered the linguistic poverty of English, he brought in new descriptive terms from Spanish, French, even native Hawaiian, and scrapped stock Alpine analogies for striking allusions to architectural forms, especially those of the Orient. The geological significance of the Colorado canyons that Newberry had proclaimed with a stratigraphic column, Dutton matched with an aesthetic canon, a stele of appreciation.

Standing on the brink of the Aquarius Plateau, Dutton exclaimed that the scene should be described "in blank verse and painted on canvas," for here "the geologist finds himself a poet." Unlike the Major, he did not command nature to stand front and center; he cajoled it with reason, arranged it with a keen sense for its wholeness, and critiqued its artful outcomes. His prose was equally firm and varied, full of suppleness and infinitely varied rhythms, an ideal medium for a landscape full of exotic shapes and surprises yet comprehensible. He differed from Powell too in that he saw the morphology of scenery by analogy to art rather than to paleontology. He found an aesthetic meaning where Powell promoted a taxonomic order. He differed from Gilbert in that he measured the landscape with figures of speech rather than with numbers and reached for architectural rather than mechanical metaphors. But what organized his influential studies of the Colorado Plateau, as with his other investigations, was a commanding sense of history, one suffused with a literary sensibility.[31]

Dutton's survey of the Colorado Plateau consisted of three monographs. Each work had a common theme, erosion; a common subject, vulcanism; and a common sensibility, the artistry of geologic time. In the *Report on the Geology of the High Plateaus of Utah* (1879–80), which consolidated Dutton's reconnaissances for the Powell Survey, vulcanism dominates. In fact, a major cause for the elevation of the plateaus amid a landscape otherwise devastated by erosion was precisely their thick volcanic caps. In the last of the triptych, *Mount Taylor and the Zuñi Plateau* (1886), Dutton portrays an inverse scene. Here a region formerly inundated by volcanic extrusions has so wasted away from erosion that only the vents and conduits that brought basalt to the surface remain, like broken ruins or toppled idols left from a vanished empire. In both studies Dutton presents his material in the form of journeys—not strict itineraries, but an imaginative recasting of events and scenes so that in the course of the journey one travels through the geologic history of the region.

Between these two studies lies the masterpiece, *Tertiary History of the Grand Cañon District* (1882). Vulcanism is a secondary feature, a decorative vignette to the great text of earth time. Here erosion and uplift exist in magnificent equilibrium; the Grand Canyon is their dynamic outcome. Remove the river, and only a plateau would remain. Remove the plateau, and there would exist only an undistinguished river valley. The two had to cross to produce rim and river. That dialectic was aptly expressed in the geologic history of the larger region.

But what made the Grand Canyon a "great innovation in our modern ideas of scenery," as Dutton declares, was another interaction, that of science with art. The sense of gigantic proportion displayed, the remarkable clarity of forms, the almost baroque splendor carved out of grand symmetries went beyond the explanatory models by which cliffs receded and rivers carved; they express the aesthetics of erosion. River and Canyon worked against each other like chisel and marble, brush and canvas. Nature was artist as well as engineer. For Dutton, unlike the members of the Ives Expedition, the scenic cliff and the geologic cliff were the same cliff. Both could be understood by similar acts of critical imagination, and one could be used to reinforce the other. The shape of the land involved mind as well as rock, reason together with impression.[32]

So journeys proceed not only through geologic time, as revealed through the stratigraphic column, but through a hierarchy of aesthetic effects, as manifest by the Canyon country's many features. Typically, a chapter of science alternates with a chapter of art. In a way, Dutton's construction of imaginary journeys by which to arrange the significant facts is not so different from Thomas Moran's invention of imaginary vantage points from which to assemble assorted features into a proper impression. With Moran, however, art in all its painterly excesses can run rampant into a glorious chromolithograph of Emerson's "transparent eyeball." With Dutton science disciplines the art so that each counters the other.

And so that, equally, each enhances the other. Aesthetics amplifies the often inadequate geology. "I have in many places departed from the severe ascetic style which has become conventional in scientific monographs," Dutton confesses. "Perhaps no apology is called for. Under ordinary circumstances the ascetic discipline is necessary. Give the imagination an inch and it is apt to take an ell, and the fundamental requirement of scientific method—accuracy of statement—is imperiled. But in the Grand Cañon district there is no such danger. The stimulants which are demoralizing elsewhere are necessary here to exalt the mind sufficiently to comprehend the sublimity of the subjects." The Grand Canyon was a scenic innovation. It required a new eye, a new voice, and a new perspective.[33]

That perspective begins on the summit of the Great Rock Staircase, the stepped cliffs that form the receding flank of the High Plateaus. The *Tertiary History* thus opens where the *High Plateaus* concludes. Here Dutton's persona surveys the scene, establishes its boundaries, sketches the whole into which each particular must fit. As Dutton's quest proceeds, journey by journey, to its climax, those individual phenomena will become more heavily laden with meaning, richer in complexity, and animated by the geographic and historical ensemble of which they are a part. But first the proportions, first the standards by which to understand and appreciate: Dutton's descent down the Great Rock Staircase to Kanab defines a stratigraphy of the landscape, a standard by which to measure both geology and aesthetics. Just as it functioned as a baseline for the Powell Survey's cartography, so Kanab serves as a kind of narrative referent for the journeys to follow.

This first traverse also introduces the informing process for the whole of the Grand Cañon district, erosion. Two images of erosion frame the text, one current, one buried in the geologic past. The first is what Dutton calls the Great Denudation. At one time the High Plateaus had extended far south, covering even the Canyon in thousands of feet of rock. Then, like a receding shore, the plateaus had ebbed to their present conditions. The reason is erosion, erosion on so vast a scale that it beggars the imagination, and with it Dutton commences the *Tertiary History*. A second, complementary image, the Great Unconformity, will conclude the monograph. Buried in the depths of the Inner Gorge, the Great Unconformity records a dramatic break in the rock record where Precambrian mountains—two ranges, one after the other—had worn away like mastodon molars to their roots, yielding a plain as level as a lake. The great stratigraphic column visible in the Canyon then arose layer by layer upon this scoured surface. Thus between them these two events define the borders of the Canyon's

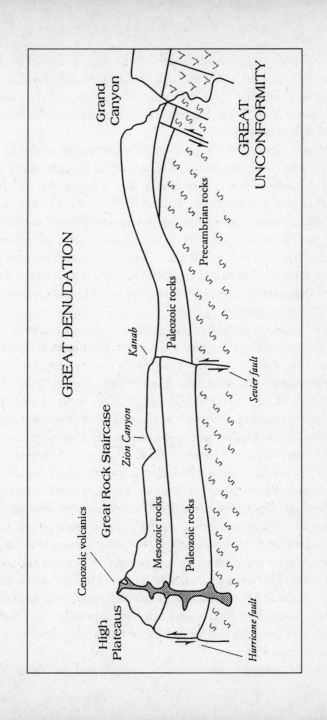

geologic history, the one its basement, the other its roof. They provide the historical heads and tails that Dutton demanded, as the
regional panorama from the Aquarius Plateau and the Canyon
panorama from Point Sublime furnish geographic boundaries.
They define the whole.

In comparison with such events the excavation of the Grand
Canyon proper is almost trivial. The Canyon appears magnificent
only because it has occurred at a time when water and rock,
downcutting and uplift, are balanced, after the river has carved
and begun widening its gorge but before new layers of rock can be
stripped away to empty sky. The middle displays what the extremes have eliminated. At the Canyon's temporal borders geologic time has abolished geologic history. Within the Canyon
itself, however, the record is writ large on lithic parchment with
the quill of erosion.

From Kanab, Dutton conducts three imaginary journeys. Each
highlights a geologic process and a scenic phenomenon characteristic of the canyon country. Each merits double chapters, one of
science, another of aesthetics. Each builds upon the other. Each
concludes with a climactic illustration in the form of a synthetic
panorama, coming as though to the end of an imaginative syllo-
gism. En route Dutton accents certain facts of special importance,
scrutinizes "type" formations and features, and constantly relo-
cates his position within the larger scope of the region. Amid a
landscape of such sweeping vistas and chaos of exotic forms, such
a typology acts to focus attention, a necessary organizing technique to provide something for the eye and mind to rest upon.
The narrative never loses shape by the sheer accumulation of fact
or overstrains itself searching for perspective. Science and art are
not hopelessly homogenized, as in the worst of German Romanticism, or diluted of presence, as in much of American transcendentalism. Instead aesthetics and science, impression and fact—a
passage of one followed by passage of the other—develop in calculated counterpoint.

The first journey travels to Zion Canyon and the Valley of the

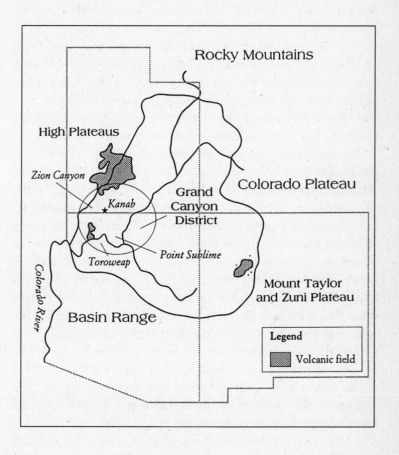

Rocky Mountains

High Plateaus

Zion Canyon

Kanab

Colorado Plateau

Grand
Canyon
District

Point Sublime

Toroweap

Mount Taylor
and Zuni Plateau

Colorado River

Basin Range

Legend

Volcanic field

Virgin, a study in cliff facades, the sculpting of buttes, and canyons carved out of the plateaus of the Great Rock Staircase. The second travels to the western Grand Canyon, specifically to that marvelous intersection of vulcanism, faulting, and fluvial erosion exhibited at the Uinkaret Plateau's Toroweap Valley. This is a study in the gorge, where one looks down rather than, as at Zion, up. But when he approaches the brink, Dutton separates himself from nearly everyone who has crept, strolled, or driven to its edge because he resists the temptation to stare down the precipice. He steps back. He measures his distances, checks to reestablish his position. Only then does he peer over the edge.

The third journey winds over the Kaibab Plateau to the eastern Grand Canyon's Point Sublime, which he named. Here Dutton announces his syntheses. The Grand Canyon, he exclaims, "is the sublimest thing on earth. It is not alone by virtue of its magnitude, but by virtue of its whole—its *ensemble.*" All the critical features of all the journeys here converge. Dutton arranges those forms, defines their regional focus, then in two magnificent chapters summarizes the geologic history of the Canyon and the evolution of a scenic day. True to form, he draws back, places Sublime amid other rim viewpoints, discusses the geologic mechanics behind the sculpting. He elaborates on Powell's theory of cliff recession and Gilbert's axiomatic discussion of fluvial erosion in the *Henry Mountains.* But these are subplots, a denouement to the story.[34]

For a story it is. There is science here in abundance, its facts and names the raw stuff of the narrative and its great laws the mechanism of plot. Yet the informing logic is aesthetic and literary. The *Tertiary History* is, above all, a geologic saga, an earth epic. Dutton had grown up in a century that discovered and exploited historicism. In natural science this meant Darwin and evolution, Laplace and the nebular hypothesis, Clausius and Kelvin and the second law of thermodynamics. The age of the earth, the age of life, the question of origins and ends all preoccupied intellectuals. But it meant also those who tried to make a science of history—Georg Wilhelm Friedrich Hegel, Auguste Comte, Karl

Marx, Herbert Spencer, to name a few. In America the century boasted William Prescott, Francis Parkman, Henry Adams, Frederick Jackson Turner, and Lewis Henry Morgan, to cite some of Dutton's more prominent contemporaries, all of whom credited history as an explanatory method and a medium of immense synthesis.

Into their ranks, on the strength of his heroic biography of the Grand Canyon, came Clarence Dutton. In an age that tried to cast its thoughts into evolutionary sequences, directed by single principles, and synthesized in grand teleological climaxes, Dutton did precisely that in the *Tertiary History*. The Grand Canyon symbolized earth history as nowhere else on the planet, dramatized the temporal ether that suffused every event, and elucidated the principles by which such a universe might be comprehended. An era that invented geology, that was in fact obsessed with the larger questions that the discovery of earth and time had exhumed like dinosaur bones found in the Grand Canyon an eloquent emblem of that mystery and in the *Tertiary History* an epic retelling of its story. The *Tertiary History of the Grand Cañon District* created the view from the rim. Together with Powell's river-defining *Exploration*, it made the Canyon truly Grand.

It was a timely no less than time-synthesizing book. First published as part of the Geological Survey's 1881 *Annual Report*, it was republished, complete with a chromatic *Atlas* in 1882, as the first of the USGS's distinguished series of monographs, the inaugural big book of arguably the most significant scientific bureau of nineteenth-century America. While mining engineer Frank Emmons might grumble that he started Dutton's book "but came to the conclusion that life was too short," reviewers like Archibald Geikie in Britain and James Dwight Dana at Yale University received it enthusiastically. It quickly established itself as a model of modern natural history.[35]

The *Tertiary History* was received in fact much like Osmond Fischer's first synthesis of the planet's natural philosophy, *Physics of the Earth's Crust*, which was published contemporaneously

with it, which quoted Dutton frequently, and which Dutton reviewed in the *American Journal of Science*. Fischer's text was an attempt to relate geophysical topics to the evolution of the globe
made possible by the contractional hypothesis. But timing also
merged natural philosophy with moral philosophy. Appropriately,
the *Tertiary History* appeared the same year that Herbert Spencer
made his triumphal tour of the United States. Simply substitute
"fluvial erosion," in the one case, and "contraction," in the other,
for Spencer's "struggle for existence" to discover that Dutton and
Fischer approximated the famous Spencerian epitome of evolution as a progression from incoherent homogeneity to increasingly
coherent heterogeneity. From the vague sweep of the Aquarius
Plateau, Dutton's narrative had advanced, stage by stage, to a
grand climax, an intricate integration of interdependent pieces, at
Point Sublime. So the Canyon's evolution underwrote the belief
that similar principles of progress informed all aspects of the
universe. That earth history could progress foreshadowed—
destined—an epoch of social progress. That such an exotic spectacle could be reduced to precepts common to prevailing scholarship validated the robustness of those principles. The Canyon
gave as well as got.

Dutton's career did not end on the rim. As Point Sublime was the
climax but not the conclusion to the *Tertiary History*, so the *Tertiary History* was an apex but not a end to Dutton's oeuvre.

   Even before he departed the Colorado Plateau—in fact as part
of his reconnaissance of the High Plateaus—he had supplied some
of the scientific muscle to bolster Powell's famous *Arid Lands* report (1878). His volcanic researches continued in Nicaragua and
Hawaii. His inquiry into the Charleston earthquake of 1886
brought international acclaim. When Powell successfully lobbied
in 1888 for an Irrigation Survey within the Geological Survey—an
institution that would direct much of western settlement if Powell
had his way—he appointed Dutton chief engineer. The assignment

was to last only two years. In 1890 a new chief of ordnance appeared who was less sympathetic toward Dutton's annual furlough to the civilians. In 1890, too, Congress began a serious attack on Powell's more geopolitical ambitions, and the Irrigation Survey folded. During congressional hearings Dutton admitted that the topographic maps on which Powell placed such hope, and on which he had pinned the future of the survey, were not suitable for reclamation engineering. The maps, Dutton insisted, were a good investment; they were not, however, essential or suitable for irrigation. Powell never forgave him.

But there were changes within the Geological Survey as well. When he first took to the field, Dutton had high hopes. "I cherish the belief that the Survey in the course of a few years will accomplish much for geology," he wrote Geikie in 1880. "The spirit which has been infused into it is most gratifying. . . . Mr. King is more than justifying all the high expectations which attended his appointment & his skill & ability to organize & administer have proved to be of the highest order. He succeeds in everything. . . ." It was Clarence King who placed Dutton in charge of the Division of the Colorado after the consolidation of the surveys, and King who supported his work at the Grand Canyon.[36]

By 1885, however, at the end of his years on the Colorado Plateau and after nearly four years of the Powell administration, Dutton had other thoughts for Geikie. "Our Survey is now at its zenith & I prophesy its decline. The 'organization' is rapidly 'perfecting,' i.e., more clerks, more rules, more red tape, less freedom of movement, less discretion on the part of the geologists & less outturn of scientific product. This is inevitable. It is the law of nature & can no more be stopped than the growth & decadence of the human body." But it was not a future calculated to inspire a man like Dutton.[37]

When he left the Geological Survey in 1890, he found himself in "virtual banishment" at the arsenal in San Antonio, Texas. He pursued some scientific work, but not single-mindedly and then with difficulty after illness broke his famously robust health. Still, Dutton so impressed George Breckenridge, a San Antonio banker, that the two went on a world cruise. There was a manuscript on China too, though it was never published. In 1899 Dutton and Breckenridge decided to tour Mexico, and for this Dutton invited his Canyon comrades William Holmes and G. K. Gilbert. "Shall be delighted to see you once more," Dutton wrote his favorite artist, "and recall old times when we were young and beautiful and when the roses bloomed—or rather when the coyotes howled and the cactus spines got into our shins." They visited ruins and volcanoes and spent the long hours on the train playing cribbage, far from coyotes and cacti. Holmes sketched, Gilbert jotted stray observations into his omnipresent notebooks, and Dutton puffed on his cigars.[38]

Later a new chief of ordnance rehabilitated Dutton's Army career and returned him as his personal assistant to Washington, D.C. There Dutton delivered his final scientific paper, on radioactivity and volcanoes, and there in 1912 he died. Perhaps apocryphally, but surely appropriately, his last words recalled his "old friends on the Geological Survey." His years on the plateaus were those he remembered most vividly, and for them history has best preserved his memory. When he arrived, the Canyon was one of many natural marvels clamoring for celebrity. When he left, it stood alone.[39]

In the end, the *Tertiary History*—and the Grand Canyon—were about seeing. Graphic art was as indispensable as Dutton's graphic prose. Its illustrations, especially its *Atlas*, are primary reasons why the *Tertiary History* has endured.

Begin with the photography. The Wheeler Survey had priority here. Timothy O'Sullivan made the first photographs of the Canyon, from the river, and after he departed, William Bell extended the view to include the rim of the western Canyon. Both, however, sank into the general morass of that troubled survey. By then Powell had introduced two other photographers, E. O. Beaman to the river, and J. K. Hillers to the rim. Their photo portfolios, done under impossibly laborious conditions, made the Grand Canyon one of the first western spetacles to receive a thorough photographic inquiry, the Wheeler and Powell survey pictures coming at nearly the same time as the photographs of Yellowstone taken by W. H. Jackson for the Hayden Survey.

Reputations were made here. For his work with the King and Wheeler surveys, O'Sullivan achieved deserved fame. For Hillers the Canyon portfolio launched him on a celebrated career that made him chief photographer for both the Bureau of American Ethnology and the U.S. Geological Survey and earned him international renown for developing techniques to render photographic transparencies on glass. In the *Tertiary History* the heliotype reproduction of Hillers's photos is poor, but the photos were included because photography was too important a visual medium to ignore and because like Dutton's habit of continually relocating himself on the landscape, they sharpen by contrast the illustrations of Thomas Moran and William Holmes.

Probably, however, their real significance was subtextual. Hillers's photographs were taken back to studios and studies where they provided, for the others, a visual archive. They complemented field sketches; they furnished a frequently consulted reference. From them Moran refreshed and reconstructed his memory. Almost certainly the best of Hillers's work was co-opted into Dutton's prose, Thomas Moran's colors, and William Holmes's indestructible lines.

The ensemble's artist qua artist was Thomas Moran. He came to the *Tertiary History* with a glowing tradition of Second Age artist-explorers to sustain him and with plenty of Canyon art already behind him.

An English emigrant from a family of painters and etchers, Moran was born in 1837, a year after Emerson's *Nature* and Cole's *The Oxbow*. He evolved as an artist by fusing the two and by investing in the natural scenes of the American West something of the landscape aesthetics of J. M. W. Turner, whom Moran admired, and the Düsseldorf school, whose American practitioners often rivaled him. Moran had Turner's eye for color, Albert Bierstadt's exuberance for the monumental, Frederick Church's passion for wild landscapes—and a nationalist's delight in the scenery of the country in which he grew up. "That there is a nationalism in art needs no proof," Moran asserted. An artist "should paint his own land."[40]

So Thomas Moran did. In 1871 he had joined the Hayden Survey to the Yellowstone. From that summer's experience came a portfolio of dazzling watercolors of geysers and bubbling hot springs and the magnificent panorama, *Grand Canyon of the Yellowstone*, an immediate sensation. Congress purchased the work for ten thousand dollars, hung it in the Capitol, and proceeded under its radiant majesty to authorize Yellowstone as the first of America's national parks. Not one to pass by publicity, Powell brought Moran to the Grand Canyon in 1873, and from the North Rim, Moran painted a companion piece, *Chasm of the Colorado*, also bought by Congress and hung next to its Yellowstone cognate. Later both traveled to the Centennial Exhibition in Philadelphia where they boasted the scenic splendor of the American empire. Powell further convinced Moran to illustrate his *Exploration of the Colorado River*, a collusion of purple prose and chromatic landscape, though Moran had to mute his vision into engravings. Alone among Canyon artists, Moran had to convey both rim and river. When Dutton began his research on the *Tertiary History*, he naturally looked to Moran, and in 1880 Moran returned to the Canyon with Dutton. It would not be his last trip.

Though in his early years he often lived by his engravings, his instincts were those of the painter for whom Art transcends Nature. "I place no value upon literal transcripts from Nature. My general scope is not realistic; all my tendencies are toward idealization." Though "of course" all art must come "through Nature," Moran believed "that a place, as a place, has no value in itself for the artist only so far as it furnished the material from which to construct a picture." In particular, "topography in art is valueless."[41]

Rather, art expressed emotion; representational art was intrinsically inferior to the "impression produced by nature" on the painter. Moran freely, promiscuously rearranged landscape features so that their positions would be true to their "pictorial," not their topographic nature. Citing Turner, Moran declaimed that "Art is not Nature; an aggregation of ten thousand facts may add nothing to a picture, but be rather the destruction of it." Literal truth counted for nothing; representation was "never a work of art, is never a picture." The artist was free to rearrange features as he saw fit. His only ambition must be to express a region's "character"; his only standard, his success at "sacrificing the literal truth of the parts to the higher truth of the whole." That Morgan did on a grand scale. The *Chasm of the Colorado* was landscape as opera.[42]

For a graphic artist, the Canyon presented special problems, technical as well as aesthetic. Dutton recognized two: the lack of perspective and the false perspective. The rim offered no simple pictorial focus. No peak, waterfall, or single gorge concentrates the eye, while a blank horizon inscribes a vast plane. In the *Grand Canyon of the Yellowstone* Moran could include the falls and gorge of the Yellowstone River within a contained vantage point. The Grand Canyon of the Colorado, from the rim, allowed no such perspective. The false perspective, a "defect which usually mars all canyon scenery," was particularly pronounced at Powell Plateau, where Moran sited his *Chasm*. It involved "the flattening of objects through want of gradations in tones and shades, and the obscurity of form and detail produced by the great distances and hazey atmosphere."[43]

But Moran's philosophy of landscape art allowed him, demanded of him, that he invent his own perspective. He constructed artificial platforms for introduced viewers and contrived foregrounds that forced the eye toward the picture's center. He orchestrated individual terrain features, repositioned them, exaggerated them, according to the overall "impression" he sought. He distinguished separate objects by their radiant colors, unmarred by sun or shade. He unified whole panoramas by an interstitial medium of cloud and haze. The obscurity of form and detail that Dutton lamented Moran turned to positive effect. "This Grand Cañon," Moran explained, "offers a new and comparatively untrodden field for pictorial interpretation, and only awaits the men of original thoughts and ideas to prove to their countrymen that we possess a land of beauty and grandeur with which no other can compare."[44]

Most fundamentally Moran devised ways to transfigure canyons into a range of mountains. By falsely lowering the foregound, he introduced a focus; by flooding the upper canvas with cloud and peaks, he erased the vast panorama of sky that filled the natural horizon. Improbably the two perspectives, rim and river, so geographically and aesthetically at odds, here combine. Instead of viewing the Grand Canyon from the top, looking down, the viewer of a Moran Canyon looks up. The river becomes a focal gorge; the inner Canyon vaguely resembles a Yosemite Valley viewed from afar. On the North Rim he had stalled and needed the prod of Powell's enthusiasm to kick-start his own. In a sense he also transferred Powell's river-dominated perspective to the rim, a move that agreed as well with conventions of landscape art and that merged with his successful illustrations for Powell's *Exploration*. The rim becomes a kind of valley that peers into a gorge behind which lies a braided tangle of other gorges.

Moran's was a transcendentalist landscape. Emerson's insistence that words were signs of natural facts, and natural facts only signs of spiritual facts, found perfect expression in a Moran Canyon. Paint replaces words. The meticulously drawn foreground, so detailed a viewer can identify lichen species on limestone boulders, substitutes for the realm of natural facts. The cloud-filled background of soaring summits and light becomes the world of spiritual fact, no longer lashed to the laws or limitations of topo-

graphic or geologic reality but rearranged to convey the larger
truths and sublimity of Nature and unified by ubiquitous spirit
clouds. The first nature of things becomes a second nature of
spirit. So too the eye moves from Canyon gloom to Canyon glory,
from Powell's Great Unknown at the center to an illuminated
panorama of sun-splashed peaks that constitute the rim and then
moves up and beyond with an arc of rainbow.

Everything in fact is exaggerated, in part to fit the Canyon into
conventions of landscape art and in part to celebrate nature on a
grand and unexpected scale. It was his "Big Picture," Moran confessed, big in ambition as well as size (seven by twelve feet). It
swept aside all previous Canyon images. It shrinks Egloffstein's
Big Cañon into a cameo, overwhelmed by a larger geographic and
aesthetic perspective. Like nearly all paintings of the Canyon, it
shows at once too much and too little. Moran could not capture
all of the Canyon, though he tried heroically; still less could he do
it by projecting the rim view as though it were seen simultaneously
from the river in a kind of Romantic cubism. Yet for all its distortions, the fantasy of its vantage point, its geography of misty marvels, no one would mistake a Moran Canyon for anything but the
Grand Canyon. And for all its bombast, the *Chasm of the Colorado* marks as much the effective onset of Canyon art as Powell's
Exploration does Canyon literature.

The *Grand Canyon of the Yellowstone* touched a more instinc-
tive chord. Mountains, not canyons, still dominated popular aesthetics. The first of the Great Survey leaders, Clarence King, made
his reputation by climbing Mount Whitney, by discovering a
glacier (the first in America) on the slopes of Mount Shasta, and
by writing *Moutaineering in the Sierra Nevada*. Forested Yellowstone, ringed with mountains, could be set aside as a "pleasuring
ground" for the people of the United States. The Grand Canyon
was more problematic. As Dutton recognized, it required special
cultivation. It remained preeminently a scientific spectacle, perhaps a nationalist emblem, not a source of popular entertainment,
easy recognition, or public endorsement. Appreciation of Canyon

scenery did not grow organically either out of folk contact or from existing traditions. It was created—had its meaning invented almost in defiance of convention—by an extraordinary elite. That act of will is reflected in the heroic contrivance of Moran's *Chasm*.

Yet he returned; the *Chasm of the Colorado* was only a prologue. Though his career began at Yellowstone, it ended at the Grand Canyon. Moran revisited the Canyon in 1880 with Dutton, courtesy of the Geological Survey; in 1892 with William H. Jackson, the great photographer of the West; and in 1901 with a coterie of artists from New York. In the latter two cases he came as the guest of the Santa Fe Railroad. By 1904 Moran was undertaking almost annual visits to the Canyon rim. A senior member of the National Academy of Design and a charter member of the Society of Men Who Paint the West, honored as the dean of American landscape painters and increasingly popularized by chromolithographs of his Canyon canvases published by the Santa Fe in an effort to publicize its Arizona route as a tourist mecca, Moran in his later years became obsessed with the Canyon. The Canyon was color—that was its maddening aesthetic challenge—but it was nature too, and it was American, uniquely so. The Canyon's rim was American art's greatest gallery and its greatest pulpit.

Moran expressed this best when he returned to the South Rim in 1901. He had always insisted that the "business of the great painter should be the representation of great scenes in nature." The grand spectacles of nature lay in America: from here too should come the world's grandest art. The one American artist to earn John Ruskin's praise, Moran now led the fight against foreign, usually French, influences—against impressionism and later cubism, against the celebration of new subjects other than landscape, against the spirit of expatriation. Such art was "false to nature" and such artists, by ignoring their proper heritage, were falsifying their future. More and more Moran's podium was the Canyon rim, a scene without any European equivalent.[45]

When the National Gallery of Art staged an exhibition on national parks in 1917, four of the forty-five paintings were Moran's, and largely on the strength of his example, seventeen of the collection were of the Grand Canyon. For his catalytic work at Yellowstone, the National Park Service proclaimed him the "father of the park system." Two years later, assisted by the exhibition, the Grand Canyon finally achieved national park status itself. Surely Moran's paintings had contributed. What Moran had fixed in paint was now fixed in law. The scene was finally preserved in statute more enduring than varnish and gilt. The director of the gallery, William Holmes himself, was no less lavish in his praise. Moran, he declared, is "the greatest landscape painter the world has ever known."[46]

Yet success had its price. By substituting an idealized landscape for the actual one, Moran had subordinated the topographic Canyon to reigning cultural ideas, aesthetic conventions, and painterly mannerisms. It may be that some of Moran's use of mist reflects the impact of Oriental art, much as Dutton (with his allusion to Oriental pagodas and Fujiyama) shows the influence of Japan in particular. But more likely Moran was simply indulging in Turnerian conventions and in the idealization of nature that, particularly in landscape art, survived the onslaught of Darwinism, realism, and naturalism, much as its techniques survived impressionism. That is what made his imagery recognizable and what made his Canyon "art." But it also bound the Moran Canyon to the beliefs and norms of those times. However prolonged their tenure, they could not survive indefinitely. The "harmonious" whole that Moran sought—the social significance of his landscape as art—would be lost when those ideas faded. The Canyon as landscape feature might, through law, be preserved, but the Canyon as landscape art could not.

"Those who have long and carefully studied the Grand Cañon of the Colorado," Dutton wrote, "do not hesitate for a moment to pronounce it by the most sublime of all earthly spectacles." Thomas Moran had studied it longer and more carefully than any other artist. The idealization of Nature that he embodied persisted in both genteel and wilder forms—in Robert Underwood Johnson and John Muir, William Keith and Albert Bierstadt. Increasingly, however, it became itself conventional. In Moran's later Canyon paintings, perspective becomes formalized, the Romantic genteel, and the operatic decadent. So stylized did Moran's Canyon become that he took to signing his paintings with a thumbprint to frustrate forgeries. His repaintings had the quality of rite; his art, not the Canyon, had become their object. From his rim ramparts Moran fought a vain rearguard action against avant-garde modernism.[47]

What he lost, ironically, was the Canyon, the real Canyon of rock, strata, open sky, silty streams, and indestructible shapes. Art, it appears, was short; life, long. The topographic Canyon that Moran dismissed as irrelevant continued long after the landscape art that he practiced morphed into other genres or passed away. What Moran lacked was the capacity to see those stones in a way outside his idealized vision. In particular he lacked a science disciplined by topography. Yet unlike Yellowstone, where art would shape science, at the Grand Canyon science would instruct art. The meaning of the Canyon depended on its scientific, not its artistic, significance. An art disconnected from those geologic values would fade with artistic fads. Dutton's disciplined text is a literary fugue, with contrapuntal science and art; unchecked, Moran's canvases swell into an opera of almost Wagnerian excess.

If not the discipline of science, art needed, at a minimum, the particulars of place. The aging Moran lost both, if he had ever possessed the former. His Canyons became the scene of painterly excess, gilt-edged relics from another time, like the split-twig figurines recovered from Paleo hunter caves in the Redwall; historical curiosities like Casteñada's and Ives's misreadings of the Canyon's future. Without a believable transcendentalism, Moran's "harmonious whole" looked as quaint as Victorian bric-a-brac, as relevant as Emerson's *Nature* in an age of automobiles and quantum mechanics. The capacity of the Canyon to surprise was gone. Enduring Canyon art would be art that stayed close to the stone.

Paradoxically, some of the most durable of Canyon art came from the man who had praised Moran as the greatest of all landscape artists.

William Henry Holmes was an ideal foil to Thomas Moran. He stands to Moran as Dutton does to Powell. Among them the four men mark the ordinal points of the heroic age; they contain the compass of the Canyon's meaning. If Dutton fused aesthetics with geology, Holmes bonded science to art.

He was in many ways typical of those untypical travelers from high culture who set out for the High Plateaus in the 1870s. He was a master of the panorama; a craftsman with line, recalling the horizontal linearity that typified much of American painting at mid-century; and a self-taught naturalist, who specialized in the material artifacts of geology and archaeology. But most important, he was a splendid specimen of the artist as traveler, illustrator, and reporter so typical of the Second Age. Holmes's panoramic landscapes, fusing enthusiasm with exactitude, belong with Audubon's encyclopedia of wildlife, George Catlin's portfolio of Indians, and Frederick Catherwood's inventory of Mayan ruins. At the Grand Canyon a man who specialized in representational art met a landscape that needed only full-scale representation.

Formally trained in art, Holmes was recruited as an illustrator for the Hayden Survey and sent west from 1872 to 1877. His keen eye soon made him a competent geologist and, from his experiences on the Colorado Plateau, a fine archaeologist, attracted particularly to the hard shape of landforms and artifacts. Later he worked for the Geological Survey, then transferred into the Bureau of American Ethnology. From there he spent three years as curator of anthropology at the Field Museum in Chicago before returning to the Smithsonian Institution, where he remained for twenty-three years and continued his passion for exploration by directing the Armour Expedition to the Yucatán. When Powell died in 1902, Holmes succeeded him as director of the Bureau of

American Ethnology. His *Handbook of Aboriginal American Antiquities* (1919) immediately established itself as the standard reference. A remarkable career in all, yet it was art that glued the various shards together, art, his first love, as William Goetzmann notes, "to which he always in some fashion returned." Appropriately his administrative tenure ended in 1920, when at age seventy-four, he became director of the National Gallery of Art. He remained for twelve years.[48]

Rehired by the Geological Survey, Holmes went to the Grand Canyon in 1880 to assist Dutton. Like his other professional work, his drawings were functional. In the *Atlas* to the *Tertiary History*, sandwiched between geologic maps, the panoramas give texture where the maps, lacking a suitable topographic contour base, show only the symbolic color of geologic eras and the gross structure of faults and uplift. The pictures give visual texture as well to Dutton's prose. What makes the *Tertiary History* more than a collection of sketches is what also makes the *Handbook of Aboriginal American Antiquities* more than the riprap of pothunters and the impressions of dilettantes in search of the picturesque: it communicates ideas as well as images. The same principles of natural history that inform Dutton's text shape Holmes's mighty panoramas as well.

A Moran landscape is a study in grandiloquence; every Moran painting has every Canyon element in it. Holmes, however, evolves the great out of the tiny. He gathers individual scenes as though they were rock samples, saved for later assembly into a composite stratigraphic column or, in the case of the *Tertiary History*, a composite history. The cliff, the gorge, the butte—all are separately featured, then combined into a succession of syntheses at Zion Canyon, Toroweap Overlook, and Point Sublime. Consider each in turn.

For Zion Holmes drew *The Temples and Towers of the Virgen*. Reinforcing Dutton's architectural analogies, Zion Canyon looms out of the desert like a cityscape, reminiscent of Frederick Church's panoramas of Greek antiquities and the Holy Land. The rock strata and land sculpture gather lines into a focus with almost mathematical perspective. It is as though the land explained itself, as if it needed only to be discovered like a fossil ammonite and recorded. Probably the panorama's most remarkable feature is the immense scale and emptiness of its sky, which consumes more than half the scene. One effect of course is to prevent a vertical exaggeration of the buttes, to shape panoramas much wider than they are high. But another, perhaps more significant consequence is that the picture illustrates one of the themes of Dutton's text, the concept of a Great Denudation, a period of erosion that stripped off the greater proportion of the rocks that had once covered the region and of which the isolated buttes and receding cliffs are but a relic. The principal scientific significance in the scene is precisely those missing strata.

Consider, next, *The Grand Cañon at the Foot of the Toroweap*, a study in fluvial downcutting, dominated by the excavations of the western Canyon. Even here, however, there are countervailing horizontal planes with which to balance the vertiginous gorge. One is the horizon; the other, the red shelf of the Esplanade, a great erosional terrace between the rim and the gorge. The geologic text is that the Canyon eroded in three stages: first the Great Denudation, which stripped away the overlaying strata; then a lesser downcutting to (and horizontal recession along) the Esplanade; and finally a renewed burst of downcutting to the inner gorge. The proportioning of the panorama records these facts precisely. The quarter or so of the panorama committed to sky retains the lessons of the Great Denudation. The Esplanade then balances the gorge. The river appears only as a patch, as though stone curtains had parted to reveal it. Matching it is a pond on the Esplanade, a reminder that water has been the sculptor of both. Still, the gorge is the most active feature, and it explodes at the viewer. Quite possibly, *The Grand Cañon at the Foot of the Toroweap* is the most successful of all Canyon art, surely the finest synthesis of rim and river.

Holmes's most ambitious pieces, however, are the three panoramas—east, south, and west—from Point Sublime, which portray the eastern Canyon and collectively illustrate the climax to Dutton's text. Here all the processes and forms of land sculpturing and all the geologic themes of the Canyon converge. Here Dutton declared was the Grand Ensemble. And here Holmes demonstrated his mastery of Canyon art. He broke the point's 270-degree sweep into three parts. Only one, the view south, includes the river, though again only as a symbolic emblem. The vast sculpting of the Canyon through the Kaibab Plateau was not the result of a river-cut gorge but an immense process of excavation by retreating cliffs, a modern exemplar of what must have occurred during the Great Denudation. Every feature is included, each in its proper place, each stratum distinct and identifiable.

The most instructive feature, however, is the foreground of the *Panorama from Point Sublime* (looking east). Two figures reside. One sketches, obviously an artist, while the other, instructing over his shoulder, is surely a geologist. Holmes and Dutton—or equally the two personas of William Henry Holmes. Viewing the panorama from the Aquarius Plateau, Dutton once declared that the geologist must find himself a poet. At Point Sublime the reverse has occurred; the artist has found himself a geologist. What made the Grand Canyon a significant landscape for Western civilization was its scientific lessons. It came too late and was too much sui generis to reform western art. Even as Moran and Holmes painted, landscape art had crossed the summit and was descending down the lee side of Western aesthetics. But in 1880 it spoke with the clarity of a Holmes line and the vitality of Moran's colors—and with the cultural authority of geologic science. Dutton the poet met Holmes the geologist.

Yet the Holmes panorama was far more than simple photographic realism inflated to an enormous scale. It had artistry and necessary distortions. No one can see line and structure in the Canyon with the clarity with which Holmes drew them. Canyon haze is real, and though Dutton warned that it could deceive, it is as fundamental to the scene as Canyon colors, with which Holmes also dispenses, or Canyon shadows, which are equally absent. Even perspective in the Point Sublime panoramas does not rely on a gathering of linear movement into a single focus, so much as it presents a foreground of curves that tend toward a planar horizon. In fact, the scene inverts traditional perspective in which an open foreground rises to a focal object. Instead, horizon takes the place of foreground, and a descending feature of Canyon topography, like a butte or gorge, substitutes for the focal figure. The eye flows naturally into the depths. Holmes neatly conveys what Moran had to contort. Every portion of the Canyon is seemingly exposed with equal exactness and transparency. One has the illusion of peering into, not merely at, the panorama, as if the picture actually exfoliated toward the viewer, opening in all directions to expose more of the scene than would actually be possible to see. Much as Dutton's word panoramas enlarge the unaided mind, so Holmes's panoramas instruct the unaided eye.

The contrast with Moran is enlightening. As befits their shared arts education, their rough compositions are often comparable. Holmes blocked out *The Temples and Towers of the Virgen*, for example, much the same way Moran did the *Chasm*. But Holmes savored the horizontal, willing to let the eye sweep across the horizon, while Moran demanded a vertical thrust, from gorge to cloud, an ascent from Canyon gloom to Canyon glory. Where Moran drew meticulous foregrounds that extend to progressively misty horizons, Holmes sketched vague foregrounds that lead to greater detail in the distance. A Moran horizon is filled with clouds and reaches to the top of the canvas; the upper half of a Holmes horizon is often empty sky. Where Moran exploits color to distinguish objects and relies on mist to organize them, Holmes uses line. Moran overwhelms; Holmes instructs. Behind Moran lay a transcendentalist philosophy; behind Holmes, commonsense realism. Whether at the rim of the Grand Canyon or at the ruins of Oaxaca, whether dealing with a vanished culture or a relic mountain range, Holmes was attracted to its material stuff—its artifacts, antiquities, rock strata, peaks. The topographic landscape that Moran denigrated Holmes celebrated. At the Canyon illustration, informed by science, became art.

Each too suffered the vice of his virtue; each exaggerated. Moran's was an exaggeration of things added, of features inserted, of colors heightened, of painterly flourish. Holmes exaggerated by removing. He achieved a false clarity by substracting unnecessary attributes, stripping away features not essential to a geologic description, like a physicist denying all attributes to an object except its mass and motion. Missing are color, shadow, cloud, haze, all of which are as fundamental to the seen Canyon as its gorges and strata. The outcome is a landscape whose emotive power derives from the scientific ideas behind it. Where a Moran landscape denies any presence but his own, the Holmes panorama boasts a deceptive emptiness that has allowed later generations to see in the scene what they wish.

William Holmes did not return to the Canyon to brood over it as Thomas Moran did or to relive past triumphs in the way Powell did. Instead, like Dutton, he finished his statement with the *Tertiary History*. Nor did Holmes feel the need to urge American scenes on to those who would produce an American art. Again unlike Moran, he worked naturally in a long tradition of American illustrators and exploring artists who emphasized reportorial fact rather than the psychological impression; who preferred line over color; who promoted a commonsense realism (through often a Romantic realism) over painterly imagination. Enchanted by ever-subtle hues and dissolving mists, mesmerized by Turnerian aesthetics, Moran could paint and repaint the same Canyon scene—was compelled to paint it over and over—and he saw his reputation rise and fall with the ebb and flow of painterly conventions. Praised by Ruskin, he was ignored by Apollinaire. Holmes drew once. There was no ambiguity about either source or significance.

William Henry Holmes could be stubborn and showed plenty of ego in his scientific work. His dogmatism while director of the Bureau of American Ethnology retarded for decades the study of early humans in the New World. But as an illustrator he consistently subordinated his talent to the subject at hand, to whatever landscape or field of inquiry he applied his art. As an artist he remained, like so much of his published work in government documents, relatively invisible to the public, shaping popular culture through his influence on high culture. Fame and power he found in the institutions of the Washington establishment; popular acclaim followed from his design and curatorial labors—the western exhibits at the Philadelphia Centennial and Columbian Exposition, the ethnographic collections at the Smithsonian, the art of the National Gallery.

And, one should add, with his enduring interpretation of the Grand Canyon. His panoramas lifted the *Tertiary History* far beyond the usual character of government monographs. They made its *Atlas* one of the magnificent books of western American art. More than Moran's *Chasm*, but like his *Grand Canyon of the Yellowstone*, they made plausible the assertion that the Canyon deserved protected status. When the National Park Service wanted to rally sentiment behind the park system and to incorporate the Grand Canyon within that realm, it was William Holmes, then director, who staged a major exhibition of western paintings at the National Gallery of Art. To help make the Canyon a park was by then easy. The real triumph had come four decades earlier, when he had made it Grand.

LEAVE IT AS IT IS: ONE OF THE GREAT SIGHTS

In less than twenty-five years American civilization had gone to the brink of the Canyon rim, ridden the rapids of its informing river, and reclassified the place from an "unprofitable locale" to the "sublimest thing on earth." What the Ives Expedition began, the U.S. Geological Survey completed. With Thomas Moran and William Henry Holmes, Timothy O'Sullivan and Jack Hillers, the Canyon entered American art. With John Wesley Powell and

Clarence Dutton, it inspired classics of western American literature. Together with John Strong Newberry and Grove Karl Gilbert they made the Canyon a centerpiece of American geology, and its larger setting, the arid lands, the focus for public land reform. Through its revelations of geologic time, the Canyon entered a larger intellectual discourse regarding the order of the universe and humanity's place within it.

Geologists were at the core, not only because the Canyon was a singularly geologic spectacle but because geologic time, like the world sea, touched the shorelines of all the great questions of the culture. They navigated that sea with the conceptual instruments of historical discourse. With laws of evolutionary progress, Newberry had joined geology to biological and theological thought. Powell demonstrated how, by similar means, the broader questions of the earth coincided with social and philosophical schemes. In practical terms he endeavored to adapt American social, political, and economic institutions to the environmental conditions of the West. Gilbert's achievement was to pioneer in the effort to unify geology by analogy to mechanics and mathematics. It was Dutton's role to fuse geology to the large questions of chemistry, geophysics, and landscape aesthetics. By assuming that evolution affected every geological phenomenon, he could relate the development of landforms, the petrological sequence of volcanic extrusions, and the history of the earth. His practical contribution was to adapt aesthetic theory, especially that common to architecture and painting, to an appreciation of the new scenery. After a hundred years the larger interpretation of the plateau province is still the one that Powell, Gilbert, Moran, Holmes, and Dutton collectively gave to it.

Yet the Canyon's assimilation into American society remained skewed. Revelation, appreciation, comprehension—all had derived from the cosmopolitan culture of a mobile elite, not from long residence or out of folk traditions. Settlers had shunned the place, as Casteñada and Ives had predicted they would, as fur trappers and frontier guides had labored to see that they did.

Even those most indomitable of intermountain pioneers, the Mormons, had done little more than run summer sheep onto the Kaibab. No one lived within the gorge.

Almost alone an educated elite had seen merit in the Canyon's bizarre land sculpture, recognized that the river intersected important intellectual questions, and sought out the rim in defiance of popular taste, Gilded Age mores, and laissez-faire politics. The High Plateaus were an overlook for high culture; the Colorado's canyons, a traverse through the cultural preoccupations of the era's intelligentsia. Yet the cultural Canyon remained incomplete. The heroic age had cut the gorge, as it were; the widening of its cliffs fell to a second generation of intellectuals and a miscellany of frontier folk, from miners to tourists, who slowly gathered into the widening gyre of the Canyon's call.

Utilization was not far behind appreciation. Its principal intellectual proponent, Robert Brewster Stanton, extended the Romantic imagination to engineering with the proposal (which became an obsession) to bring a railroad through the inner gorge. Stanton came to the project with nearly two decades of practical experience in the West and several construction triumphs to his credit, a reputation as a brilliant young engineer, and an imagination inflamed by Powell's narrative about what Stanton ever after referred to as "the River."

Though Stanton later downplayed the epic scale of the project—insisting that he was merely putting numbers on land features—the scheme had become for him, as his earlier expedition had for Ives, a dream, a vision of linking America's greatest landscape feature with its mightiest technology. In fact Stanton's 1889–90 trips down the Colorado gorges were a logical successor to Ives's investigation of the Colorado's navigability for steamboats. They established Stanton as a pioneer among river runners and marked the return of the engineer as explorer. The historical researches that he began in 1906, though never published in full, made Stanton, along with Frederick Dellenbaugh, the foremost early historian of the Colorado River.

Yet little practical came from it. The tragic death of Frank Brown, his sponsor, that marred Stanton's first expedition, the collapse of his railroad project, the popular triumph of the Powell and Dellenbaugh chronicles—all put Stanton's Canyon reputation, and that of Canyon engineering, into a recession from which it has never fully recovered. Not until 1923 was a real engineering survey of the Canyon conducted, and then for reservoir sites. Not until the 1960s was another engineering scheme on this scale proposed, only to face a more furious public skepticism than even Stanton had endured. Not its economic but its intellectual resources continued to shape the interpretation of the Canyon and the politics of its development.

That meant geology.

The mantle fell first to Charles Doolittle Walcott. Walcott had studied for a while under Louis Agassiz at Harvard University, then worked for James Hall at the New York State Museum, soon distinguishing himself as a paleontologist and stratigrapher. When the Geological Survey was formed in 1879, Clarence King quickly hired him. He promptly joined Dutton for stratigraphic work on the High Plateaus, and two years later he went into the Grand Canyon itself. Powell led the assault in person, carving a horse trail down into the Nankoweap Basin. Walcott recalled how "encamped in snow, often concealed for days in the driving frozen mist and whirling snow, the party gradually overcame the apparently insurmountable obstacles in its way. . . ." The result was a "perfectly frightful trail," Dutton informed Geikie.[49]

The excitement over, Powell departed, and Walcott overwintered, working out with painstaking care the stratigraphy of the Grand Canyon and Chuar groups. It was entirely in keeping with Powell's temperament that with great gusto and in defiance of winter storms, he should blaze a trail into a remote region and then leave the real scientific work to another. He had done it with Gilbert; he did it with Dutton. And it is entirely indicative of Charles Walcott that he would follow Powell's path and with meticulous care make its possibilities known. Whether he realized it or not, in taking this promising protégé to the place at which he had made his own reputation, Powell was setting Walcott on an administrative career that would far surpass his own.

While he showed talent as a scientist, Walcott demonstrated genius as a politician of science and an administrator of scientific organizations. Within a year after he had emerged from the Nankoweap Trail, Walcott was promoted to the post of chief paleontologist for the Geological Survey and almost immediately thereafter he succeeded a weary Gilbert as chief geologist. He began his Grand Canyon investigations shortly after Powell took charge of the survey; when he concluded them in 1895, he had replaced Powell as survey director and was entering a fabulously successful career as the chief administrator of half a dozen bureaus and the instigator of half a dozen more. But his scientific reputation, upon which those appointments first depended, stemmed from five papers on Canyon geology that he published between 1880 and 1895.

Walcott rebuilt the USGS after Powell had plunged it to political depths with the kind of reckless abandon he had shown in constructing the Nankoweap Trail. Later he applied the same patient hand to the development of other institutions: the Bureau of Mines, the Carnegie Institution of Washington with its Geophysical Laboratory, the National Research Council, the National Advisory Committee on Aeronautics, the Smithsonian Institution and National Museum of Natural History, and the National Academy of Sciences, of which he was president. For a while in 1907 Walcott actually administered three organizations simultaneously: the Geological Survey, the Reclamation Service, and the Smithsonian Institution. As quiet as Powell was flamboyant, Walcott translated ideas into institutions. That the U.S. Geological Survey became the prototype of the Progressive Era technocratic bureau, that as "the Mother of Bureaus" it spawned new offspring almost annually, was largely a tribute to Charles Doolittle Walcott.

What he got from the Canyon, he returned. His behind-the-scenes politicking was instrumental in promoting the forest reserve program, under the provisions of which the Grand Canyon was first set aside as protected land, and he was active in the drive to establish the National Park Service, under whose rule the Canyon has been administered since 1919. Through his investigations in the basins of the Nankoweap, the Kwagunt, and Chuar, Walcott worked out the stratigraphy of the eastern Canyon, thus complementing the stratigraphic profile of the western Canyon published by Newberry and Gilbert; recommended theories of structural and fluvial history for the evolution of the Canyon, modifying and challenging the simple model Powell had proposed; and correlated Canyon strata to the historical geology of North America, a much-needed modernization of Newberry's pioneering studies. Eventually he carried that stratigraphy, and the evolutionary perspective it archived, far beyond the Colorado Plateau.

Most famously Walcott mined the Burgess shale of the Canadian Rockies for fossils that, better than anywhere else, mapped out the explosion of life that had characterized the Cambrian era. A century later, after another round of scientific revolutions in biology and geology, Walcott's evolutionary vision could seem archaic. But when the earth's age was believed to be less than one hundred million years, when the texture of fossil-preserved evolution remained thin, those excavations had come as a revelation. The scientific reputation that Walcott inaugurated at the Canyon, he confirmed in the Rockies.

Nothing endures forever, ideas no less than things. An evolutionist like Charles Walcott, a man who commenced his career in the gorges of the Grand Canyon, knew that all too well. He had seen the evidence both in the rocks he had cored for fossils and in the more impressive testimony of erosion that had stripped whole strata away, ripping out great chapters from the text of time. The cairns along the Nankoweap Trail resembled so many geodetic markers by which to triangulate into unknown basins of earth time, and for Walcott they continued that task long after the talus of new discoveries had buried his old track. Behind the Burgess shale stood the Hermit, Bright Angel, and Hakatai shales of the Nankoweap and Chuar basins; behind his vision of evolutionary history stood the whole stratigraphic column of the Canyon like an immense stele. Not his conception of evolution but the evidentiary fossils he gathered would endure; not his ideas but the institutions he collected, packed, and stored as he did his trilobites. For Charles Walcott that was likely enough.

Time also had its shape. The stratigraphy of the Canyon was one dimension; the sculpturing of those strata another. Explaining the geomorphology of Grand fell largely to William Morris Davis and his student François Matthes.

The first systematizer of physiography (geomorphology, as it later became known), Davis dominated the science for half a century. He publicized as well as consolidated the Powell Survey's insights, cultivated his reputation on theirs, made the place and the people fundamental to his endless proselytizing on behalf of the American school of earth science. He wrote biographical memoirs of both Powell and Gilbert for the National Academy of Sciences; Gilbert's was the largest in the series. And he spent nearly two decades propagandizing about the Canyon country as a paradigm of fluvial erosion.

He summed his insights in the concept of a geographical cycle. This was a model that synthesized geologic structure, erosional processes, and boundless time into regular patterns of land sculpture and, especially, of landscape evolution. Every climate, every geomorphic province had its distinctive variants, but each played out a universal trend. Landscapes were closed systems that eroded: they began as uplifted, high-energy systems and then decayed through the entropy of erosion into level, low-energy systems. Landscapes proceeded through cycles in the same thermodynamic sense that heat engines did. The process was irreversible, though it was possible to restart the cycle with renewed uplift. Thus landscapes were a kind of entropy clock, and sculpted landforms, a variety of earth fossil. Recognize the pattern, and you know a landscape's stage of evolution; you know its age.

This was an extraordinary premise. The geologic vision that stratigraphy had built up, piling stratum upon stratum in chronological ascension, the geographical cycle metaphorically tore down, methodically stripping away layer upon layer. Instead of stratigraphic columns, Davis proposed a series of systematic declinations by which, for every imaginable climate, uplifted landscapes wore away until they culminated in a nearly level erosional baseline he termed a peneplain. Erosion thus produced a parallel chronology to deposition. Landscapes became geomorphic fossils that tracked the tread of geologic time. It was possible to date the earth even when its preserved records were being removed, not a trivial achievement, given that most of geologic time was lost to erosion's erasure. Davis proposed means to measure that void.

For a geology of denudation rather than of deposition, there were few spectacles to rival the Grand Canyon. In the early twentieth century William Morris Davis went there more than once to see for himself the fluvial mechanics that Gilbert had discovered, to inspect the Canyon terraces that Dutton had recommended as marking epochs of erosion, and to ponder the meaning of the Great Denudation taking place along the distant cliffs and of the Great Unconformity, which mapped an ancient peneplain deep in the gorge. All the critical parts were here, and so was the medium for their synthesis. From the rim Davis marveled at the supremacy of geologic time over all things, even the Grand Canyon. "The most emphatic lesson that the canyon teaches is that it is not a very old feature of the earth's surface, but a very modern one; that it does not mark the accomplishment of a great task of earth sculpture, but only the beginning of such a task; and that in spite of its great dimension, it is properly described as a

young valley." In brief, Davis played Dryden to Dutton's Shakespeare, pruning and systematizing, reducing the captain's aesthetic flights of geopoetry to the heroic couplets of an inspired
pedagogue.[50]

Wherever the Davisian model traveled, it did so with the authority of the Colorado Plateau and the Grand Canyon behind it.
As he had made the Powell Survey fundamental to the historiography of the American school, so Davis made the Canyon indispensable as the symbol of its lessons. He first wrote of the Canyon
in 1892, relying on the published studies of the Powell group. But
between 1900 and 1904 he led parties of American geologists to
the rim four times, and in 1912, under his inspiration, the American Transcontinental Excursion organized by the American Geographical Society established the Canyon as a centerpiece in its
travels.

Meanwhile Davis wrote and spoke avidly about the significance
of the Canyon—"my crack piece," he called his stock lecture, one
he tirelessly propagated throughout the world. He helped make
the Grand Canyon as familiar to Europeans as the Alps were to
Americans.[51]

But even before Davis began his famous European campaign on
behalf of the precepts of the American school, Europeans were
themselves coming to the rim. Karl Baedeker included a map of
the Canyon and instructions on how to reach it by stagecoach in
his 1899 edition of *The United States*. Buffalo Bill Cody escorted
British nobles to the North Rim in a vain effort to have them purchase the land as a private hunting preserve. Baedeker's approach
to the Canyon as commodity proved more successful, and among
the many spectators arriving at the rims were geologists. Some
came as members of international congresses; others, as private
pilgrims; most, as skeptics regarding the grandiloquent reports of
their American colleagues.

For Archibald Geikie the intriguing precepts of American

fluvialism appeared to be singularly dependent on the observations freighted back from the Colorado Plateau, and he finally went west to see for himself. He was not disappointed. "It is unquestionably true that had the birthplace of geology lain on the west side of the Rocky Mountains," he wrote to critics of fluvialism, "this controversy would never have arisen." Dutton had said much the same, though, with respect to the impact the Canyon might have made on Western art. It had not influenced European art as much as Fujiyama had Japanese art, but it did so influence Western science. Geikie, who had reviewed Dutton's *Tertiary History* for *Nature*, brought just that message back to Europe. As his point of departure he compared the "gorges of the Old World and the New."[52]

The Canyon became to geology what the Louvre was to art or St. Peter's Square to architecture. In 1891 it was thus wholly appropriate that Powell and Gilbert should direct the International Congress of Geologists to the rim and that in 1904, a part of the St. Louis fair, the World's Geological Congress should arrange a special excursion as well. For perhaps fifty years afterward, however, it was not necessary to bring the Muhammads of the earth science to the Canyon; the lessons of the Canyon were brought to them in abundance and without further contention. It had become an effective symbol of American geology: its exploring traditions, its conceptual lessons, its relationship to American nationalism. But it had also become a symbol of the earth, available to all.

Science dominated Canyon culture, and geology ruled Canyon science. When C. Hart Merriam, then director of the Biological Survey, came to the region in 1899, he was more interested in the stratification of life zones he saw in the San Francisco Peaks than in the stratification of evolutionary ages in the Canyon. For this the Canyon may perhaps be grateful. Merriam's life zone concept was almost pure Humboldt, practically a plagiarism of outmoded

ideas Humboldt had developed for the slopes of Mount Chimborazo in Ecuador.

From anthropology, however, the Canyon may have given as
much as it got. Its meager human ruins were dwarfed by the magnificent cliff palaces at Mesa Verde, Chaco, and Canyon de Chelly;
its inner Canyon residents, the Havasupai, proved consistently
less interesting than the well-ceremonialed Hopi; the outlying
Paiute and Hualapai claimed the fringe of anthropology as they
did the margins of the Great Basin.

Yet the Canyon influenced the careers of at least two men who
did much to shape American anthropology, in both its ideas and
its institutions. Powell became the first director of the Bureau of
American Ethnology in the Smithsonian, and William Henry
Holmes succeeded him. The BAE later sent ethnographers back
to the scenes of the founders, while the Havasupai, as they had
time and again since Spanish times, attracted others. Meanwhile
its discovered ruins, however diminutive by regional standards,
did provide the political pretext for President Theodore Roosevelt to invoke the Antiquities Act and set the area aside for special protection within the public domain.

So also Canyon art remained subordinated to Canyon science.
The Canyon never became a distinct genre, the way the Sierras or
the Hudson River did; it contributed nothing by way of new techniques or a new oeuvre. The Canyon had to compete with more
traditional Alpine landscapes, and even Thomas Moran remained
popularly acclaimed as an "artist of the mountains." It had to
compete too with that other genre of western painting, the West
as action. Charles Russell, Frederic Remington, and their imitators—none went to the Canyon to paint. They needed human figures, not sculpted buttes, and human drama, not the inexorable
patience of geologic time. They demanded events, not scenery.

By the time Moran, assisted by free passes from the Santa Fe
Railroad, returned and made the rim into a pulpit, landscape was

fast fading from the forefront of art, an avant-garde was poised to replace the artist-explorer, and even Remington was turning to impressionism. More and more the Canyon became the symbol of past glories, not an emblem of new ideas. On its rim Moran found a great constant with which to preserve American traditions that he thought representative. It seemed that artists might paint the Canyon as they would the old masters in the Louvre or the classic scenes of picturesque Italy, as a form of instruction by imitation. The Canyon received rather than inspired.

Perhaps there was no need for anything further. In 1902 the torch effectively passed. Frederick Dellenbaugh, the seventeen-year-old artist on Powell's second voyage, published a comprehensive history, *The Romance of the Colorado River*. For its frontispiece it had a watercolor sketch from Thomas Moran, who a year before had made the visit to the Canyon rim that led to his American art essay; for its epilogue, the book included a biography of John Wesley Powell, who died shortly before publication. The *Romance* became a kind of panegyric to Powell, and Powell before his death urged Dellenbaugh to complete the saga with the full narrative of the 1871–72 voyage, a task Dellenbaugh finished a few years later. Together the books consolidated the Powell epic, transformed a popular saga into scholarship, and brought the Canyon into the canon of American history.

But if the *Romance* summarized the region's heroic past, the *Grand Canyon of Arizona*—a compendium of photographs and word pictures about the Canyon promulgated that same year by the Santa Fe Railroad—pointed to its future. Filled with original essays by Powell, Moran, Stanton, and R. D. Salisbury of the Yale Geology Department, the book also contained poetic bromides from writers like Harriet Monroe, overwrought effusions from professional lecturers like John L. Stoddard, and trite apostrophes from recycled hacks like Joaquin Miller and Charles Dudley Warner, all of it intertwined indiscriminately like the cross-bedding of the Coconino sandstone. The Santa Fe wanted tourists, needed to fill up the otherwise empty miles of rail across the great plateaus and justify the spur track constructed to the South Rim and the luxury hotel at its terminus, and publicized art as a means to transform the Canyon into a tourist mecca. The railroad sought a wide audience, not necessarily a discerning one. The result is an anthology of some of the best and some of the worst of Canyon literature, a symbolically transitional work.

There was more. John Muir published his own Canyon meditations, siting the scenic Canyon among the Wild West's other marvels. What had been difficult to reach was now easy, he noted. Muir, however, appreciated that the real question of access was mental, and it consisted of words or of concepts or perspectives by which to shape words. The Grand Canyon was not simply larger than other features—a score of Yosemites could nestle among its ravines—but far more alien. Yet if words failed, the conceptual apparatus did not. The Canyon's very distinctiveness had become an asset, and its assimilation far advanced. So also in 1902 François Matthes began reducing the topographic exotica of the eastern Canyon to cartographic exactitude by pounding in the initial benchmarks for triangulation, the first automobile, a steamer, drove from Flagstaff to the South Rim, and the Kolb brothers, future impresarios of mass culture, arrived at the rim. An era of heroic invention passed into one of popular dissemination.

The following year President Theodore Roosevelt journeyed to the rim to see for himself this fabled "natural wonder, which . . . is in its kind absolutely unparalleled throughout the rest of the world." At the Canyon the great American nationalist—an accomplished naturalist, pioneer conservationist, author of *The Winning of the West*, a wildly popular politician—met a great American scene. "Leave it as it is," he urged. "The ages have been at work on it, and man can only mar it. What you can do is to keep it for your children, your children's children and for all who come after you, as one of the great sights which every American, if he can travel at all, should see." American exceptionalism had found

America's most exceptional landscape and made it a national icon.[53]

Among the last of America's landscapes to be formally explored, the Grand Canyon had become among the first of its natural marvels and, for a nation that tended to substitute natural monuments for cultural ones, entered the pantheon of its sacred places. Its valorization offered as much a cross section through American history as of earth history. The evolution of that interpretation had, with eerie symmetry, mimicked the evolution of the Canyon's features. The spasmodic tectonism of geographic exploration, the varied tributaries that flowed into the main currents of American thought—with breathtaking brevity the two processes had met, merged, and not merely laid down a course of history but entrenched it so deeply that the Canyon became a permanent feature of America's cultural landscape.

CANYON
AND COSMOS

No place, not even a wild place, is a place until it has had that human attention that at its highest reach we call poetry.

—WALLACE STEGNER

Even after 100 years, however, the explanation of this landscape still defies us.

—CHARLES HUNT

Before the nineteenth century had ended, the Grand Canyon had found its poet. It also had its troubadours, its court painters, its cultural viziers. The achievements of that exploring elite, whose reputations had grown symbiotically with the Canyon's, had shattered its intellectual and aesthetic isolation, had smelted its unearthly novelty into prevailing genres, had hammered its common stone into the statuary of a national icon. The Canyon was not merely assimilated by American culture; it was smothered, polished, celebrated. A fearsome landscape had become a precious object.

What ideas had valorized, politics moved to preserve. In 1893 President Harrison included the eastern Canyon within a forest reserve. In 1906 President Roosevelt proclaimed it a game reserve and two years later a national monument, thanks to its scattered Indian ruins and its significance to geologic science. In 1919, after extensive lobbying, Grand Canyon became a national park under the jurisdiction of the recently (1916) established National Park Service. What began as a symbol significant to a cultural elite became a pleasuring ground for the public. And more: the Canyon was transfigured into a sacred site. Like Hindu pilgrims in a *yatra* of Shaivite shrines, Americans (and foreign visitors in search of America) toured the national parks, of which the Grand Canyon was the most distinctive, the one without comparison anywhere else. The Canyon put America First.

Interest was sufficient for a spur track from the main Atchison, Topeka, and Santa Fe railway line to replace the stagecoach in 1901, and the luxurious El Tovar Hotel, a legacy of tents and rustic cabins. The rails brought the Canyon into the realm of industrial tourism. Ease of geographic access soon demanded an equal ease of intellectual access. Powell had pronounced the Canyon the most sublime of earthly phenomena but one that demanded patience and toil; Dutton had insisted that the scene could be appreciated only through strenuous cultivation, the hard labor of thinking. Neither suited the Santa Fe Railroad, the National Park Service, or the touring public. The visitor replaced the explorer, the Kodak snapshot the grand canvas, the inscribed overlook the monographs, atlases, and personal narratives of its bold creators. Visitation shifted from the lofty North Rim to the more proximate South. The Grand Canyon was fast becoming a commercial commodity and a cultural cliché. And there were fewer intellectuals of stature to say otherwise.

Even as the public crowded Canyon overlooks, intellectuals were walking away. The new high culture of modernism had little use for the High Plateaus. Modernism busied itself with other projects, above all itself. It preferred to search its own depths, not those of the river-excavated Kaibab and Uinkaret plateaus. Scenes and values that had previously joined elite to folk were worn away, a Great Denudation within American culture. Modernism cut through their once-common ground until intellectuals and the public were as distant and incommensurable as the two rims. A commercial popular culture rushed in to fill the void.

MODERNISM MOVES ON: THE POPULIST CANYON

Still, a few of that cultural elite came to the Canyon, and they filled the yet empty places with books, paintings, scientific essays, even exploring sorties. But in this silver age the emphasis was on refinement of data and technique. The larger vision had been established; the remaining chore was to inventory, document, fill in, and, above all, to disseminate. High culture had to compete with purveyors of the Canyon as experience, not simply as idea; with the Canyon as a promotional medium, not the expression of scientific information. The genius of the age was to bring the viewer to the Canyon, not the Canyon to the inquirer in the form of print or painting or principle. In literature the trend was almost wholly toward tourist books or sketches from the perspective of the tourist. Canyon oracles were those specimens of local color, like John Hance who affected the pose of frontier sage or William Bass with his homespun doggerel. They were all figures who could entertain as well as guide, who could, with a little professional promotion, become objects of interest in themselves.

So the cultural chasm between elite and mass widened, partly because high culture was going elsewhere and partly because popular culture became ever more estranged from modernism. The Canyon's silver age produced the splendidly intricate topographic maps of François Matthes, the sparkling watercolors of Gunnar Widforss, the stratigraphic inquiries consummated by Edwin McKee, and the painstaking engineering survey of the

Colorado River by the U.S. Geological Survey. But the popular interpretation of the Grand Canyon passed to the promotional department of the Santa Fe Railroad, to the photographs and motion pictures of the Kolb brothers, to naturalists in the employ of the National Park Service, and especially around the turn of the century to assorted characters in residence like Bass and Hance who catered to the tourist trade. Buffalo Bill tried to flog off the North Rim as a hunting preserve to British nobility; Buffalo Jones became a national celebrity for roping mountain lions out of Kaibab pines; Uncle Jim Owens, government hunter, became a character study for literary hacks like Zane Grey; stuntmen landed biplanes on the Tonto Shelf; Tom Mix rode up Canyon trails for Hollywood's silver screen. Cultural artifacts scattered like pebbles under the tracks of the Model Ts that swarmed to the rim. The Canyon was less a national oracle than a promotional backdrop.

Yet there was more at work than raw American commercialism. The Second Great Age of Discovery had inspired the heroic age of Canyon exploration and the golden age of its interpretation. By the time Teddy Roosevelt stood on the South Rim, however, the Second Age had fled to the poles, and the perspective-framing Enlightenment was fast becoming moribund. By the time Grand Canyon acquired national park status, the Second Age had died on the Antarctic ice sheets, imperialism had fallen into exhaustion in the trenches of the Great War, and Western art and science were sucked into the maelstrom of full-blown revolution. The Canyon was caught in that cultural wind shear. The gusts that had carried an age to the Grand Canyon now blew elsewhere. Even as the tourist Canyon became ever more popular, as visitation swelled and commercial interest boomed, the intellectual Canyon tumbled down the cliffs of high culture like boulders pushed over the rim by bored sightseers.

What had previously converged on the Canyon now split. In painting and the plastic arts, impressionism, cubism, surrealism, dadaism, and their proliferating progeny introduced new conceptions of perspective, broke down representational aesthetics, insisted that truth did not reside in the object but in the viewing artist. In natural science, Newtonian physics splintered against the new worlds of nature—the microcosm of atom and gene and the macrocosm of cosmology. Quantum physics and Einsteinian relativity challenged the epistemology of Newtonian science and the orthodoxy of omniscience that the Enlightenment had constructed out of its presumptions. The sheer multiplicity of human experience, the discovery of fundamental principles of indeterminacy, and the inescapable reality of relativized perspectives—all shattered the grand design of historicism. The observer prevailed over the observed. Modernists privileged irony over purpose. For philosophy William James satirized Herbert Spencer's evolutionary synthesis in an unforgettable parody: "Evolution is a change from nowish untalkaboutable all-alikeness by continuous stick-togetherations and somethingelseifications." Make it new, Ezra Pound insisted. Modernism did, with a vengeance.[1]

The old, even the geologically old, was no longer novel, could no longer serve as a universal repository of events and ideas. Modernism prepared to raze the past as Le Corbusier proposed for Paris in order to construct his Radiant City. In literature, in philosophy, in mathematics, science, and the plastic arts, avant-gardes appeared that had little use for the techniques with which the Grand Canyon had been revealed and valorized. Intellectuals were more likely to follow Joseph Conrad into a heart of imperial darkness or Sigmund Freud into the depths of the unconscious and the labyrinth of dreams than to gaze once more with John Wesley Powell into the Great Unknown of the Inner Gorge. The formative event in American art was not the exhibition of paintings that Stephen Mather and William Henry Holmes organized at the National Gallery in 1917, much less the promotional exhibits sponsored by the Santa Fe Railroad, but the exhibition of The Eight and especially the 1913 Armory Show that brought for the first time that panoply of modernist art to the United

States. The revolutionary sciences were those that scrambled after the new physics, downgrading Isaac Newton from the rank of philosopher to engineer, and that pursued the new genetics, consigning Charles Darwin from the Sage of Down to butterfly collector. In the earth sciences the controversy that gripped geology was directed not at the 1923 Birdseye Expedition through the Colorado canyons but at the international debate three years later on the theory of continental drift.

Even with the Sante Fe and the National Park Service as patrons, even with the National Gallery willing to host exhibitions, modernists were more inclined to choose Taos or Carmel than Grand Canyon. The siren song of Mabel Luhan Dodge proved more compelling than Powell's commands; the attractions of a Taos populated with pueblo Indians more stimulating than the empty immensity of Toroweap; a colony of fellow-traveling artists more comforting than a corps of far-ranging naturalists. In practice Taos artists were more modernizers than modernists, more willing to adapt new techniques than to surrender America's commonsense realism. But their shift in style necessarily meant a shift in subject as well. The Canyon became peripheral, more a sink for sketches than a source of inspiration. Regional artists preferred the mesas and ravines of northern New Mexico, which were not only closer at hand but closer in composition to Cézanne's archetypal Mont Victoire; they preferred sunburned natives in picturesque villages to cosmopolitan scientists confronting the great questions posed by a monumental nature. Emblematically the defining Southwest novel of the era, Willa Cather's *Death Comes for the Archbishop* (1927), ends with the erection of a cathedral, not the evolution of a canyon.

The intellectual elite no longer heard the call of the Canyon because they no longer spoke to it, no longer listened for its reply. High culture left the high rims for other ideas and other places, though its legacy and many of its agents remained at the Canyon, lingering like the last refracted rays of an autumnal sunset.

* * *

Among those who did come was composer Ferde Grofé who produced the Canyon's most enduring musical composition, *Grand Canyon Suite*.

Grofé had encountered the Canyon during the course of travels in Arizona as a pianist in vaudeville houses, hotels, and dance halls. His *Suite*, begun in 1921 and completed twelve years later, typically modernized musical effects without adopting a full-blown modernist philosophy. It told the same story through sound that Dutton had with words, the evolution of a day. *Grand Canyon Suite* built logically on Grofé's earlier *Mississippi Suite*.

An "obsessed" Grofé demanded a full palette of "orchestral colors," for he not only saw but "heard" color. With its mimicry of mule and storm and sunset, the symphony created a kind of musical representationalism of the sort that had fallen out of favor with musical culture much as representational painting had with the avant-garde of the plastic arts. The suite was to music what Thomas Hart Benton's muscular landscapes and Carl Sandburg's verse were to the painting and poetry of the day, an American accommodation, an adaptation of new techniques to traditional subjects. Self-consciously Grofé sought an "American idiom" that would fall "easily on the average ear." At his best Grofé was a Gershwin for American nature.[2]

But perhaps the most telling of its contrapuntal passages was historical. By the time the *Suite* was fully performed, modernism was rewriting the software of American culture like a computer virus. An intellectual migration, fleeing fascism in Europe, reinforced that trend. Equally, however popular the Canyon's depiction in concert halls, the actual Canyon had its tourism gutted by the Great Depression and subsequent wartime rationing. Visitation stalled. The Canyon lost not only its economic patrons but its political clientele as well. Grofé's work stands virtually alone, a monadnock rising from a base-leveled cultural countryside like the Mesozoic Red Butte and, like it too, a relic from an earlier era, in this case the 1920s when most of the work was written.

The Canyon remained a supremely visual spectacle, and as long as there was patronage, painters erected their easels on the rim. But modernism posed important challenges during the 1920s, as did the Great Depression of the 1930s, and there remained the imposing presence of Thomas Moran.

Moran remained the patriarch of Canyon artists, subsidized by a Santa Fe Railroad that also reproduced his paintings as chromolithographs. From the *Chasm of the Colorado* in 1873–74 to *Grand Canyon* in 1920, his art had spanned forty-seven years. No one had pondered the Canyon longer. More and more, however, his gilt-edged, heavily varnished canvases seemed like Victorian relics from a time when art read nature for its lessons, when the larger the spectacle, the grander its significance. His idiom was set. He repeated himself as a traveling bard might retell Homer or a concert pianist replay a favorite Beethoven sonata. Yet although the aging Moran continued to harangue from his Canyon lectern, increasingly big looked like bombast. His panoramas were as empty of modernist irony as they were filled with Emersonian mist.

Others sought more appropriate stances and less grandiloquent gestures and, loosely orbiting around the Taos and Santa Fe schools, developed more suitable techniques. Louis Akin, William Leigh, Oscar Berninghaus, Gustave Baumann, and Ferdinand Burgdorff all incorporated elements of impressionism or modernist mannerisms though in ways that did not lose the ultimate presence of the Canyon itself. In the end the Canyon carried their art, not their art the Canyon. Akin lamented that there was a limited market for Canyon pictures. Revolutionary modernism looked elsewhere. Modernist art explored the foundations of painting the way modernist mathematics did the symbolic logic of numbers. The era's most celebrated meditation on time was Salvador Dali's *Persistence of Memory*; surrealism, not stratigraphy, shaped its contours. Intellectually the rift between Canyon and art widened. Geographically the Canyon was suspended between two regional colonies, the Taos-Santa Fe artists and the Southern California impressionists. It remained peripheral to both.[3]

There were two exceptions, two artists—curiously amid the vogue for American regionalism—both émigré Swedes, both born in 1879. Both painted broadly representational landscapes. There the similarities between Carl Oscar Borg and Gunnar Widforss end. For Borg the Canyon (and its surrounding plateau province) were the backdrop to a spiritual quest, one that never ceased. For Widforss the Canyon became home.

Borg had a hardscrabble childhood, salvaged by personal mysticism and a sympathetic pastor who recognized the youngster's talent for drawing. A series of misadventures took him to France and England, where he languished in poverty until he began painting sets at the Drury Lane Theatre. In 1901 he sailed to America, which he entered illegally while sequestered in a propeller shaft. Two years later he arrived in Los Angeles, found work and friends, made a reputation and contacts. Through Charles Lummis he learned about the Southwest's landscapes and native cultures, and through Eva Lummis he was introduced to patrons, most notably Phoebe Apperson Hearst, who subsequently sponsored his career and travels. In particular she arranged for him to paint the Hopi and Navajo Indians under the auspices of the University of California and Bureau of American Ethnology.

His 1916 sojourn to the Southwest was an epiphany. He immediately empathized with the natives; place and people matched perfectly his own temperament. His career boomed, further supported by work for United Artists studio as an art director and bolstered by national enthusiasm for southwestern scenes. Like most of the Taos and Santa Fe artists, he was interested primarily in the native peoples, secondarily in landscape by itself.

Landscapes served most often as backdrop, a way of placing the people. What he had done for theaters and studios, he repeated in his paintings. The land was a stage for the human presence.

Despite praise from Moran himself, who reportedly considered Borg "preeminently the highest grade of any artist in America today," Borg's Canyon paintings are curiously weak, among his most feeble. A few succeed where, as with *The Great River, Grand Canyon*, the viewer is situated at mid-level and the Colorado provides a strong, sinuous focus. But in comparison with Borg's Canyon de Chelly landscapes, his Grand Canyon seems uninspired. Perhaps the scale was too immense, the contrast between the Canyon and the human presence too great. The open sky that serves the transcendentalist Borg as mist did Moran is lost amid the Canyon's closed strata. In instances where he inserts human figures, the proportion often seems odd or the commentary almost comical, as when in *Hush of Evening* Borg has a party of Navajo horsemen peering improbably into the gorge. Without those figures, however, the canvases seem weirdly vacant, neither charged with spirit nor filled with human aspiration, but simply empty, like stage backdrops shorn of their performance.[4]

The Depression coincided with personal disappointments and dampened Borg's enthusiasms. In 1932 he wrote that he could no longer look "with longing romantic eyes at the great American desert." Art had been a "religion" for him; now the ritual rather than the spirit remained. The painting continued, though wanderlust and drink increasingly filled his days. His travels returned him to Sweden, and in 1938 he married a Swede and took up residence in Gothenburg. But he was no longer Swedish and returned to California to discover that he was not American either. He died dining alone in a Santa Barbara restaurant.[5]

The contrast with Gunnar Widforss is curiously complete. Trained as a decorative painter (with a certificate from the Royal Technical Institute in Stockholm), Widforss abandoned the business to become an artist and indulged in years of travel throughout Europe and the United States. He soon specialized particularly in landscapes and watercolors. In 1912 he began exhibiting his works, successfully. He resumed his rhythm of travel and painting, even through World War I, shifting from war-torn central Europe to its nordic perimeter.

In 1921 he traversed the United States on his way to the Orient. At Yosemite he met Steven Mather, director of the National Park Service, who persuaded him that the national parks would provide a suitable subject, and Mather adequate patronage. Widforss began to tour and paint the major parks, including Grand Canyon. By 1924 he had enough works—seventy-two in all—to sustain a one-man exhibition at the National Gallery of Art. Acclaim followed. So did other exhibitions and commissions; more travel took him to Mesa Verde and then to Taos and Santa Fe.

But the Canyon most gripped him. There he set up permanent residence, becoming a U.S. citizen in 1929. There too he produced in a series of watercolors some of the most vivid and scenically honest of all Grand Canyon art. For nearly twenty years Widforss focused his art on the Canyon, claiming in effect the mantles of Moran and Holmes. At his best no one has synthesized Canyon form and Canyon color so successfully. No less an authority than William Holmes himself declared, during a Widforss exhibition at the National Gallery, that "these are the finest things of their kind that have come out of the West . . . [by] possibly the greatest watercolorist in America today."[6]

Strong words, backed by an oeuvre of Canyon art unmatched for half a century. Widforss's field of vision is more restricted than in the macroscopic panoramas of either Holmes or Moran, but it encompasses all the essential Canyon features, and it maintains a sense of definition from foreground to horizon. In Widforss's finest works, the foreground crowds to one side, the slope of the rim matches the slash of the river, the backdrop fills with mesa or butte. There is depth without an appeal to defiles, a sense of recognition without the drama of revelation, a realism that values the observed object as much as the observing artist. With his meticulous draftsmanship Canyon shape never dissolves into mist and Canyon color reinforces that sense of definition rather than, as in the cases of Moran and the later quasi impressionists, smears it. Had he been born half a century earlier, Widforss might have become a rival to Winslow Homer; or had he directed his brush to the modern landscape of machine America, to Charles Sheeler; or had he used the Southwest scenery to delve into a psychological landscape, of Georgia O'Keeffe. Instead Widforss was too committed to representational art to contribute to modernism, and too late to join the formative era of Canyon painting.

It is reported that one of the Taos artists said of Widforss, "If he just wouldn't copy nature so closely he'd be the greatest living painter in America," and that when Widforss was told of the comment, he swept his hand toward the Canyon and asked, "Could imagination improve on that?" That was not the posture of a modernist. Besides, Canyon art—the best of it—had communicated idea as well as image, and ideas are conspicuously lacking in Widforss. Neither geologic time nor transcendentalist nature figures forth. Perhaps Widforss was constrained by his public image as "the painter of the National Parks," the promotional uses to which his painting were put, the need to connect to a touring public not enthusiastic about abstract art. Perhaps the Canyon, where he died and lies buried, was greater than his art of naive realism.[7]

Widforss painted nature less as landscape than as still life. Revealingly the human figure is absent from a Widforss watercolor. There are no awestruck travelers like those Moran and Egloffstein included, respectively, in the *Chasm of the Colorado* and *Big Cañon at Diamond River*, no emissaries of high culture inserted as Holmes did at Point Sublime. So too it is hard to place Gunnar Widforss. In the panorama of Canyon history, he is at once undeniable and invisible. Like one of the buttes he loved to paint, segregated from both rims, Gunnar Widforss stands by himself, oddly isolated from the great eras of Canyon culture. Yet unlike Carl Borg, he had found a home, and unlike even the Taos artists, excepting perhaps O'Keeffe, he is indelibly identified with the scenes of that chosen place.

* * *

There were others content, eager to detail with scientific and engineering exactitude what the heroic age had poetically envisioned. Together they moved rim and river beyond the realm of science into engineering. For each the Canyon confirmed a career and established a reputation. As Widforss had become the "painter of the national parks," so François Matthes became their cartographer. The political journey that Powell had launched from Green River and that ended with the Geological Survey, the survey continued when it dispatched Claude Birdseye to map the river following the 1922 Colorado River Compact, which apportioned the waters of the Colorado among the states within its watershed and liberated those waters for reclamation projects.

An immigrant from the Netherlands, Matthes approached the rim with an excellent technical education, supplemented by engineering studies at the Massachusetts Institute of Technology. In 1896 he joined the cartographic division of the U.S. Geological Survey, serving his apprenticeship in such rugged locales as the Bighorn Mountains of Wyoming and the Bradshaws of Arizona. Impressed, Director Walcott assigned him to the Grand Canyon, a year after the Santa Fe railhead reached the South Rim. There Matthes began to map out the Canyon's most celebrated features and so helped triangulate from its exploring past into its tourist future.

The challenge was formidable, and like Walcott, Matthes was compelled to blaze trails to the North Rim (first along Bass Canyon, then Bright Angel) as well as develop new techniques for analysis. With only an alidade and plane table as instruments, the mapping of the Bright Angel and Vishnu quadrangles became a technical triumph, a tour de force of draftsmanship that demanded a different scale of contoured line and developed concepts that were to be applied to other regions. In 1904 and 1905, moreover, Matthes studied geomorphology under William Morris Davis at Harvard, and when the Bright Angel quadrangle was published, it included on its back side a geomorphic analysis written by Matthes.

The pioneering Matthes then moved on to other parks. His maps helped do for the creation of Glacier National Park what Moran's watercolors had done for Yellowstone, and decades of study helped confirm the Sierra Nevada as a testimony to glacial erosion as the Colorado Plateau was for fluvial. Compared with the Grand Canyon, Yosemite Valley seemed geographically quaint, though its exfoliating granites and glacial-polish walls made the Canyon's angular features seem cartographically benevolent. The Bright Angel quadrangle, in particular, endured as an aesthetic and technical standard, unsurpassed until the advent of laser-based theodolites and aerial photography allowed for a revision.

The remainder of the project devolved on to Richard Evans, a Matthes protégé. Over the period 1920–23 Evans mapped most of the eastern Canyon and complemented Dutton's Oriental placenames and Matthes's Nordic ones with allusions drawn from Arthurian legendry. Lancelot Point and Excalibur joined Vishnu Temple and Wotan's Throne. By the time the survey ceased, the most complex of western landscapes had become the foremost of its mapped terrains, and the Canyon had launched still more national careers.

The Geological Survey matched rim with river. The cartographic inventory that Matthes and his associates had made for the excavated plateaus, the survey's Grand Canyon Expedition of 1923 did for the gorge of the Colorado River. Under the direction of Claude Birdseye, then its chief topographic engineer, the survey recapitulated the journey of its old charismatic chief, John Wesley Powell, but with a rigor unknown to Powell's intrepid adventurers. E. C. La Rue surveyed potential damsites. Lewis Freeman reported the adventure. Emery Kolb served as chief boatman. At its conclusion one of the boats (the *Grand*) was shipped to the Smithsonian Institution for "permanent exhibition." Such was the self-consciousness of the Canyon's silver age.

The work demanded precision, which meant it proceeded with a caution that bordered on tedium. Since the compact had made Lee's Ferry the divide between the upper and the lower basin states, the expedition launched from there and made it Mile 0, a base line for the language of topographic engineering. Then they overwrote Powell's fulsome prose with a new Canyon graphic. Previously unnamed features—rapids, bends, sandbars—were to be named by their distance from the Lee's Ferry baseline. Six Mile Wash, 104 Mile Rapids, Mile 181, a most exotic landscape, a scene that left professional poets babbling platitudes, had become among the most precisely delineated places on the planet. A gauging station at the confluence of the Colorado and Bright Angel Creek completed the mathematization of the Canyon. Not only the river's shape but its dynamics came under constant, quantitative scrutiny. Numbers replaced figures of speech. Reservoir sites (eight in all, with thirteen alternates) complemented scenic overlooks. Contour maps complemented the tradition of the panoramas with which Holmes had recorded Canyon topography.

Among them, Matthes and Evans, Dutton and Powell, and the Birdseye Expedition effectively numbered as well as named the Canyon. They made the Canyon one of the most instantly recognizable sites on earth. But despite the real hazards of travel, the virtuoso technical skills displayed, and often the intellectual achievement of these projects, they were, in a sense, a colossal infilling of exploratory trenches dug by the heroic age. They marked in blanks; they fussed over details; they cleaned up untidy corners of topography. They resolved many of the minor questions of Canyon science and answered none of the fundamentals. If anything, the detailing cluttered the founding clarity of the Canyon's importance. By the 1920s no one could say, with much confidence, just how the Canyon had formed. In fact, no one could say without irony just what the place of this supremely geological phenomenon was within geologic science.

* * *

Geology mattered.

Of all the emissaries of high culture that had elevated the Canyon into significance, none had been more important. Geology as an intellectual enterprise had made the Canyon, for an American intelligentsia and its institutions, something more than a freak of nature. Geology had posed and answered vital questions of what it meant to be human and, at places like Niagara Falls, Yellowstone, and Grand Canyon, what it meant to be an American. Geology had sponsored Canyon art, Canyon word paintings, Canyon science, Canyon politics. For its geological significance, the Supreme Court had legitimized the reservation of the Canyon under the provisions of the Antiquities Act. And the Canyon had reciprocated. It had given American earth science a most cherished symbol. Reviewing the *Tertiary History*, the British journal *Nature* had even then considered the Canyon "the grandest and most unique feature in the geology of the United States, and in which indeed there is no parallel elsewhere in the world."[8]

By the late 1920s geology and especially the American school had reached a cultural equilibrium. The long revolution had ended, replaced by professional norms and bureaucratic mores. As Dutton had prophesied, editors deleted *sublime* as they would *traprock,* excised aesthetic meditations as though they were paranormal séances, and dismantled grand ensembles into working hypotheses and annotated bibliographies. No longer cascading from newfound mountains, the stream of geologic discovery slid into a lake of governmental and academic institutionalization, depositing its data like silt into library deltas. But the full impact was not obvious at the time.

A climax came in 1928, when two giants from its formative era published their final studies: Thomas Chamberlin his *Two Solar Families*, which expanded midwestern and western geology into an evolutionary cosmology, and G. K. Gilbert his *Studies of the Basin-Range Structure*, posthumously completing the cycle of works of one of the Canyon's greatest scientific explorers. That same year Norman L. Bowen published *The Evolution of the Igneous*

Rocks, a summa of another American tradition, petrology. Bowen proposed a geochemical progression by which an undifferentiated pool of magma could successively evolve into an abundance of mineral species. There was more: radiometric dating techniques were announced that promised an absolute time scale that could resolve the age of the earth; Richard Byrd launched his expedition to Antarctica, with its flight over the South Pole, that would complete the last geographic ambition of the Second Great Age of Discovery; and Herbert Hoover, a mining engineer, had become president. From minerals to planets the American school had achieved a spectacular synthesis of historical geology centered on the concept of evolution. It was a tradition often leveraged from a fulcrum at the Grand Canyon.

It was appropriately in 1928, then, that American geologists published the proceedings of a symposium in which they rejected the concept of continental drift. Over the past thirty years modernism's storms had blown through physics, biology, painting, philosophy, astronomy, and literature. A year earlier Werner Heisenberg had announced his principle of indeterminacy; a year later P. A. M. Dirac had fused the two extremes of the new physics into relativistic quantum mechanics. The theory of continental drift, as proposed by Alfred Wegener, promised to do the same for the earth sciences. Yet American geologists, especially, resisted. They had their own heroic history, their own landscape confirmations of great ideas; they preferred to follow the Great Surveys over the Great Basin and the Sierra Nevada, to Yellowstone and Yosemite and of course Grand Canyon, rather than trudge behind Wegener over the Greenland ice sheet. Continental drift was not only untrue; it was also unnecessary.

Yet their imagined climax was, as it so often proves to be, a prideful pause before the fall. By the 1930s the American school of earth science was in serious trouble, and within another decade it was moribund, as quaint before modernist science as the American regionalism of Grant Wood, much less the still-reproduced lithographs of Thomas Moran, was to dadaism, cubism, and sur-

realism. American geology had run out of frontiers. Geologic reconnaissance went the way of free land, fenced by intellectual
equivalents to the Taylor Grazing Act. Its informing question, the
age of the earth, was close to solution. Where they once revealed
ancient geologic empires, geologists now subdivided strata with
the intensity that Padua scholastics had glossed Aristotle.

More and more often American geologists engaged in a kind of
ritualistic re-creation of past glories, resurveying the classic study
areas of their heroic age, debating endlessly the theses and concepts of the great masters, recounting over and over the sagas of
western exploration. The library replaced the field as a source of
data. Senior scientists argued questions like the origin of granite
or the stratigraphy of the deep ocean as though they were disput-
ing free will or the existence of angels, unconcerned that no empirical test or experiment could drive toward a conclusion. They
revisited the puzzle of rim and river as mathematicians would Fer-
mat's Last Theorem. The Canyon became more museum than
laboratory. It endured primarily as a symbol, no longer a dynamic
source of vital information.

Ideas and energy that would have launched an international career a half century before languished in dry dock. Despite the long
tenure of Edwin McKee—first as chief naturalist of the park
(1929–40) and then as de facto chief scientist of the Canyon
region—his classic stratigraphy of the Paleozoic Canyon went
onto library shelves rather than into the backpacks of a new generation of exploring geologists. A historical geologist like McKee
would have appreciated the importance of timing: the Canyon
could no longer make a career. The great monographs and personal narratives of the heroic age were the ammonites and pelecy-
pods of a lithified past—horizon fossils of an age that no longer
flourished. Instead McKee's magisterial studies of the Kaibab and
Redwall limestones, of the Supai and Toroweap formations, and
of the Coconino sandstone largely banked the coals of Canyon science. Even geologists still read the rocks with the texts of Powell,

Dutton, Walcott, and Davis before them. The revolution in earth science came later, and it looked elsewhere.

The dominant audience was tourism. Perhaps the most interesting of those who catered to it were the Kolb brothers, Emery and Ellsworth. Arriving at the Canyon in 1902 to "follow the work of scenic photography," they eventually traversed the Colorado River from Green River, Wyoming, to the Sea of Cortés. The bulk of the journey came in late 1911, the stretch from Needles south in 1912 and 1913. They lagged ten years behind the railroad to the rim and two years behind the first strictly tourist trip down the river (that by Julius Stone), but they systematically photographed their adventure, using both still and motion pictures, and wrote a widely distributed account, *Through the Grand Canyon from Wyoming to Mexico* (1913).[9]

The voyage was adventure, to be sure, but it was enterprise too, done so that it could be photographed. The Kolbs didn't reveal the Canyon; they used the notoriety of the Canyon to promote themselves and their work. Ellsworth confessed ingenuously that "the success of our expedition depended on our success as photographers. We could not hope to add anything of importance to the scientific and topographic knowledge of the canyons already existing; and merely to come out alive at the other end did not make a strong appeal to our vanity. We were there as scenic photographers in love with their work. . . ." A scenic photographer needed a scene; that the Canyon provided. In truth the idea of scenic photography as a career came to the Kolbs after they had arrived at the South Rim, and they began their collective career by photographing the mule trail parties that ventured down the Bright Angel Trail.[10]

Here was Everyman as explorer, describing for a nation of Everymans how they did it and telling the story through popular mediums—a motion picture and a picture book, the first really of the modern era. More than anyone else, even more than the Santa Fe Railroad, which after all brought in professional writers and painters, the Kolbs transformed the genres of high culture into the medium of popular culture. The whole package of voyage and book had been staged so as to present the story, and the Canyon filmed to satisfy the eyes of the average tourist. That they faced real dangers and that the practice of photographing from the gorge required muscle, pluck, and wit spared them from being shills of the pseudoevent. They were originals, though popular culture over the decades commercialized them into a cliché.

The timing of their trip as much as its subject determined the success of the Kolb enterprise. It was an era in which alarm at the closing of the frontier and years of sloganeering by social Darwinists had led to a vigorous search for personal adventure, by elite and folk equally. The Kolbs completed their descent through the Canyon the same year that Teddy Roosevelt delivered his invocation to the "life of strenuous endeavor," and Ellsworth's book came with an introduction by Owen Wister, a Roosevelt crony, whose novel *The Virginian* had appeared the same year the Kolbs arrived at the Canyon and whose theme was the cowboy as western folk hero.

But unlike Wister, or his friend Frederic Remington, or their mutual friend Roosevelt, the Kolbs—and their colleagues and imitators—offered adventure without the trappings of high culture. Their motion picture was followed by others, as Hollywood made the West into the western, with Tom Mix, to name one star among many, riding up canyon trails and swimming Colorado gorges. Their photographs of the tourist Canyon were, as Robert Euler has remarked, the first of an avalanche, the art form of Everyman. "In sheer numbers, there probably have been more pictures taken than words written about Grand Canyon, but not much more. . . ."[11]

The changed perspective followed a change in audience. The interpreters of the Canyon were no longer intellectuals, or the Canyon's patrons even the affluent late-nineteenth-century travelers of railroad and resort, but Everyman and his Model T. By 1926 auto travel exceeded rail. Within a few years paved roads reached to all the major overlooks, a footbridge spanned the Colorado, and a park naturalist regularly lectured eager visitors. The average stay could be measured in minutes, hours at most, enough to authenticate the visit with a souvenir and to circle the overlooks as Muslim pilgrims did the Kaaba. The Kolbs retired to their studio on the Bright Angel Trailhead, as much a tourist feature as Powell's memorial or Mather Point. There they replayed their motion picture and for decades ended as they began, by selling photo supplies to visitors and photos of mule trains to tourist tenderfeet. They had become themselves part of the scene.

All this is not to say that the coming of mass culture and its intimate ally, mass tourism, to the Grand Canyon was unimportant. Upon it was predicated the public conservation of the Canyon, and its preservation in a form that would retard the crudest forms of commercialism that had, for example, blemished Niagara Falls. Nor is it to say that those who catered to tourism saw the Canyon only over the rise of a cash register. Nor, again, does this imply that the art of Gunnar Widforss was trivial, or that Edwin McKee's studies of Paleozoic stratigraphy were merely self-indulgent, or that François Matthes and Richard Evans and Claude Birdseye were dilettantes. Their contributions were as important to the Canyon's cultural history as that history had become to the American experience.

But they were minor events in intellectual history, almost provincial in comparison with the ferment in relativistic cosmology, quantum theory, the modern synthesis of evolution and genetics, the breakthroughs in formal logic, or the stunning revolution in modern art and literature. The Canyon's cliffs were no mirror for modernism, as they had been a palette for Romantic art and a slate for natural science. No Nobel laureate began a career on rim or river. No major artist shattered old genres or announced an avant-garde manifesto among its sunset-blasted buttes. No book foamed up from its rapids to demand a place in the modernist canon. Yet just such triumphs had happened in the preceding half century.

Instead the Canyon was becoming a museum piece, a fixture like statuary on the lawn of the national manor. Purveyors of the populist Canyon promoted their wares by and large without the aid of cultural elites. Tourists came to see what they were told to see; they saw (or heard or read what they were supposed to see) without seeking to turn personal vision into cultural insight. They rushed to the El Tovar Hotel, Verkamp's, the Kolb Studios, or the Hopi House for souvenirs and entertainment as much as to the Bright Angel Trailhead or Hopi Point. They visited the Canyon as they did other sites of national pilgrimage. They saw and touched an iconic Canyon. They impressed their perspective with their sheer numbers.

The Canyon was becoming culturally moribund and, for that reason, politically vulnerable. Dutton had imagined a time when the Canyon itself succumbed to the continued ravages of the Great Denudation, and William Morris Davis had exploited that image to impress on his listeners the immensity and power of geologic time. Geologically that event, if it ever proceeded to completion, was millions of years away. Its intellectual equivalent, however, was almost at hand. The cultural Canyon had reached grade.

The Canyon remained a supreme spectacle, though of what precisely was less certain. "I have heard rumors of visitors who were disappointed," wrote J. B. Priestley in the best Canyon essay of the era. "The same people will be disappointed at the Day of Judgment." A British man of letters, Priestley was a visitor twice over, and he sensed both the strangeness of the place and its significance. "If I were an American," he concluded, "I should make my remembrance of it the final test of men, art, and policies. I should ask myself: Is this good enough to exist in the same country as the Canyon? How would I feel about this man, this kind of art, these political measures, if I were near that Rim?"[12]

The Canyon was known everywhere for being known. It was fast shedding its cultural vitality, surrendering spectacle for celebrity. Modernism had pulled the plug of its intellectual life-support system, while the Great Depression (and later wartime gasoline rationing) kept tourism on a short leash. Priestley's electric enthusiasm arced across contrasts that were as much cultural as scenic. Instead a ceremonial, vaguely pathetic quality hung over natural history studies at the Canyon. Leslie Spier's report on the Havasupai promised less a descent into the Great Unknown than a search for Shangri-La in which an impassable tangle of canyons assumed the role of Tibetan peaks, shielding a valley from the restless tentacles of industrial society. (James Hilton published *Lost Horizon* four years later.)

Most revealing, however, were expeditions mounted by the American Museum of Natural History in 1937 to two isolated mesas. One trekked to Wotan's Throne; the other, to Shiva Temple. There they sought out evidence of speciation. A journey to the Canyon's buttes could complement Darwin's voyage to the Galápagos Islands. The expedition to Shiva and Wotan was thus a delayed echo of an age that had sent ships to every Pacific reef and across Arctic ice floes. Roy Chapman Andrews, director of the American Museum of Natural History, celebrated expedition leader, a later model for Indiana Jones, proclaimed Shiva Temple "a lost world."[13]

But what time and space had made, a later time and another space could unmake. Humboldt could not have catapulted to fame had he climbed the Flint Hills of Kansas or botanized from a dugout along the Rhine. An exploring age that had fought through the cataracts of the Congo and clambered over the passes of the Tien Shan was reduced to scaling the Kaibab limestone, resupplied by airdrops in metal milk cans. Time had ebbed to a lost horizon, and space had shrunk to a Canyon mesa the size of a football stadium. The episode spiraled out of control, as the popular press transformed hope into hype and plumped what was, in reality, an elaborate backpacking trip into a search for dinosaurs and the Land Time Forgot.

In the end the expedition discovered no novelties; the same species claimed both the plateau and its mesa miniature. Nor was the expedition an innovation; it was itself a diminutive descendant of the voyage of the HMS *Beagle* and the Great United States Exploring Expedition around the world. There was scientific validity to the quest and a truth to the yearning for new frontiers and a Ulysses-like ambition "to sail beyond the sunset, and the baths of all the western stars." Such had been the stuff of the Second Great Age of Discovery; such had been the motivations of the Canyon's discoverers and the context of their revelations.

But the Second Age had entered a deepening twilight. Collecting trips for squirrels and lizards on Shiva Temple had to compete, as serious biology, with the modern synthesis of evolution and genetics, rapidly composing under such figures as Theodosius Dobzhansky and Julian Huxley, even as the museum's climbers scrambled up and down the Kaibab limestone. Public expectations had more in common with Arthur Conan Doyle's *The Lost World* than with time-sculpted buttes. A recapitulation of Darwin's insights—the Canyon's mesas as a geologic Galápagos—could not compare with the neo-Darwinian synthesis of systematics, genetics, and evolution by natural selection.

Priestley had thought that the Canyon represented "our nearest approach to fourth-dimensional scenery," that here "some elements of Time have been conjured into these immensities of Space." So they had, and so had history and geography converged to make the Canyon Grand. The Shiva Temple and Wotan's Throne expeditions stood to the Second Age as ceramic figurines in the Fred Harvey gift shop did to the wild vistas beyond the windows. The cultural collusion that had shaped the Grand

CANYON GLOOM TO CANYON GLORY

Thomas Moran, *Chasm of the Colorado*. Moran succeeded in translating Alpine art into a Canyon setting. No one more actively promoted the place as a setting for landscape art. His Canyons span forty years.

SILVER AGE: ART

Gunnar Widforss, *Powell Plateau, Grand Canyon*. Widforss at his finest, and an interesting artistic backsiting on Moran's *Chasm*, which, geographically, views this scene in reverse.

THE COMMERCIAL CANYON

ABOVE: The opening of the Sante Fe rail link between Williams and the South Rim, the beginning of wholesale tourism and the construction of modern facilities to service it, 17 September 1901.

OPPOSITE, TOP: Buffalo Bill shilling in 1892 on behalf of speculators trying to flog the North Rim off as a hunting preserve to British nobility. While rich in deer and cougars, the place was simply too remote. Here the group admires the South Rim.

OPPOSITE, BOTTOM: The Kolb brothers' studio, perched on the rim like a box camera, ready to photograph mule trains.

PRESERVING ONE OF THE GREAT SIGHTS

ABOVE: Theodore Roosevelt on mule, John Hance mounted behind him, and Ida Tarbell two farther back.

OPPOSITE, BOTTOM: Steven Mather, director of the National Park Service, at the dedication of the Powell Memorial, an event where history, myth, government, and scenic values all converged. The photograph is by Francis Farquahar, who later assembled a masterful first bibliography of the Grand Canyon and Colorado River.

THE GRAND ENSEMBLE AT THE CAPITAL

After he was assigned to the Geological Survey's national office, Gilbert and some colleagues began to stage commemorative lunches, which they called the Great Basin Mess. These evolved into elaborate affairs. Much of the Washington scientific establishment, including those whose careers were tied to the Canyon, attended. The photograph above shows Gilbert (seated, third from left); Powell (seated, front); William Holmes (seated next to Powell); Charles Walcott (seated, far right); and WJ McGee (seated, left), a Powell protégé and one of the architects of a philosophy of national conservation.

SILVER AGE: SCIENCE

OPPOSITE, TOP: Edwin McKee as a young naturalist lecturing on the South Rim, near Mather's memorial plaque.

OPPOSITE, BOTTOM: Mapping the river—the U.S. Geological Survey's Grand Canyon Expedition of 1923 under Claude Birdseye.

ABOVE: Mapping the rims—a survey team under François Matthes.

RIGHT: Dr. David White and Edwin McKee examine fossils on a canyon trail.

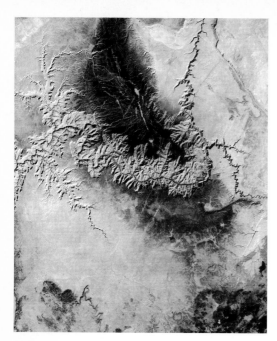

THE COSMIC CANYON

LEFT: Landsat views the Grand Canyon, reducing its immense complexity to the status of a mudcrack.

BELOW: Comparative status of the Grand Canyon and the Valle Marineris on Mars.

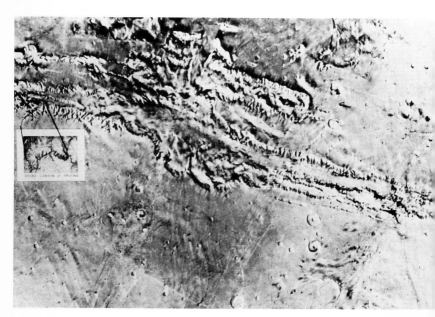

Canyon broke apart. What cultural elites sought the Canyon did not hold, and populist experiments like the Shiva Temple expedition ended in near parody. The real "lost world" was the Second Great Age of Discovery.[14]

Down the River and Back from the Brink: The Environmentalist Canyon

Yet the Canyon remained. Tourism might sag, intellectual enthusiasm falter, and the Canyon's very familiarity breed, if not contempt, then a complacency about what it signified, but the place remained, its meaning ready for rehabilitation. Just as those Colorado rapids renewed when side canyons debouched flood debris into its channel, so new interests poured into the Canyon from larger cultural watersheds and revived the vigor of its passage. The Grand Canyon entered—helped define—the white-water age of postwar environmentalism.

The first of the rejuvenating rapids was a literary revival of the heroic age.

When Wallace Stegner published *Beyond the Hundredth Meridian: John Wesley Powell and the Second Opening of the West* in 1954, his evocation had a bardic quality, a sweeping reminder of times past. Stegner claimed he was writing the "biography of a career," whatever that was. In fact, the result was a magnificent evocation of the whole Powell Survey and the landscape with which it had made a reputation. The book's "biography" was as much of a place, the Canyon Country, as it was of Powell; its narrative of a "career" was, as much as anything, a wonderful retelling of how the Canyon in particular had been explored, been named, been celebrated, and become assimilated into American culture. There were sketches of all the participants. While Powell remained at the core, Dutton earned a lengthy portrait (Stegner had written his doctoral thesis on Dutton).

The book was an immense success, an instant classic that transcended inherited categories of literary biography, history, and western literature. It immediately overwhelmed two feeble studies from the 1940s, Edwin Corle's improbably named *Listen, Bright Angel* (1946) and Roderick Peattie's anthology *The Inverted Mountains* (1948). It even found a children's cognate of sorts in Marguerite Henry's Newberry Award–winning *Brighty of the Grand Canyon* (1953). Brighty did for the Grand Canyon what Smokey Bear did for forest fire. The baby boomers who grew up with it learned that, like the burro Brighty, the Grand Canyon should remain "wild and free."

So while Brighty roamed as a free spirit, Stegner's book carried the load. Twenty years later Sally Gregory Fairfax, summarizing two decades of vigorous environmental reform, declared for an "impressive consensus" that "if you are going to read one book in this area, it should be Wallace Stegner's *Beyond the Hundredth Meridian.* . . ." If the Canyon had made Powell's reputation, his book about Powell's Canyon made Stegner's. More than anyone else Wallace Stegner became the West's man of letters and a sage for a new era of environmentalism. The heroic age had its poet. If it was through Dutton's eyes that tourists saw the Canyon's sunsets, it was with Stegner's voice that they read his words.[15]

The lesson arrived just in time. Two postwar development schemes, both promoted in 1956, proposed fundamentally to restructure both rim and river. The river's redefinition began in 1948 with the completion of the Upper Colorado River Basin Compact, which liberated the Bureau of Reclamation to plan for the wholesale reconstruction that in time made the Colorado one of the most engineered waterways in the world. Following an important controversy, the process culminated in the Upper Colorado River Storage Project Act of 1956. The rim's redefinition sprang from the National Park Service, which argued successfully for a decade-long, system-wide rehabilitation of facilities and infrastructure that it called Mission 66. If the river's reconstruction recalled the traditional economics of power and water, the rim's pointed to the recreational demands of a consumer society. Both schemes inspired an environmentalist reaction.

The political drama began slightly earlier, coincident with Stegner's *Beyond the Hundredth Meridian*. The Bureau of Reclamation had originally proposed to develop, as part of the Upper Colorado Basin Project, a dam at Echo Park that would inundate part of Dinosaur National Monument. The precedent for flooding a piece of the national park system horrified conservationists who had never quite recovered from the flooding of Yosemite's Hetch Hetchy Valley. Fighting the dam at Echo Park marked the political metamorphosis of an aging conservation agenda into a new-era environmentalism. The transformation was one of style as much as theme.

Almost all the activist apparatus applied to later controversies emerged at Echo Park. There was a picture book, *This Is Dinosaur* (1955), an anthology promoted and published by Alfred Knopf and edited by Wallace Stegner. Though it appears quaint when compared with the slick-page writing and high-voltage politics packed by its successors, the volume established a new genre. Not surprisingly the story of Powell and his Colorado adventure figured prominently. Not only was a principle at stake—the inviolability of the national parks—so was a part of American history. There was a film about Dinosaur too, another precedent. And there were hard, fascinating lessons in mobilizing public opinion to oppose large-scale development projects on the public lands. Like the old forty-niner who seemed to materialize at every western mining strike and instruct the community on how to proceed, the successful veterans of the Echo Park fight wandered widely among the public lands and controversies and provided leadership for a generation of environmental activists. In particular, dams became favorite targets for bringing public protest to critical mass.

Through those fights, through the arguments and ideas that underlay that movement, an intellectual culture began returning to Grand Canyon. Unwittingly the Bureau of Reclamation helped point the way. In return for abandoning the Echo Park site, it constructed a high dam at Glen Canyon, upstream from Lee's Ferry, and as part of its construction wave, like a slow flood down the river, it proposed hydropower dams at Marble Canyon and Bridge Canyon that would encase the flanks of the eastern Grand Canyon in reservoirs. Incredibly, the scheme even argued for a tunnel to carry waters from the Marble Canyon reservoir beneath the Kaibab Plateau and empty into Kanab Canyon, where another dam would impound it. Some 90 percent of the flow through the classic Grand Canyon would cease. Although the dams avoided the park itself and were allowed under the provisions of the park's organic act, they clearly compromised the natural integrity of the Canyon. And they thrust the Canyon center stage into American environmentalism.

Neither was the rim spared. The National Park Service itself saw to that. The Santa Fe Railroad had kept Canyon visitation on a parity with national rates during the Depression while the Roosevelt administration, through the Civilian Conservation Corps, had upgraded much of the park's facilities. Both crashed during the war. When visitation revived, tourists were trading rail for roads. The National Park Service—not only at Grand Canyon but throughout its holdings—found that its infrastructure was inadequate. In his column for *Harper's*, Bernard De Voto proposed the country abolish the parks if it wasn't willing to maintain them. The public rallied behind the parks, and the Park Service responded with a grand scheme to accommodate the automotive tourist.

The outcome was a decade-long construction intended to conclude on the fiftieth anniversary of the service's organic act. Mission 66 brought Levittown, through the interstate highway system, to the national parks, reconciled outdoor recreation with a consumer society, and ensured public access to a motoring public eager to fill the new infrastructure. Yet while it built visitor centers, it did little to explain the meaning of the parks beyond bureaucratic celebration of the Park Service. Mission 66 revived the wholesale construction projects of the CCC and coincided, historically and culturally, with the creation of Disneyland; critics charged that neither was an appropriate model for a national park. If the Bureau of Reclamation could see rivers only as damsites and kilowatts, the Park Service seemed driven by visitation rates and paved overlooks, the nature reserve as theme park.

Both schemes questioned what a park meant and especially what this particular park meant. The Echo Park controversy partially answered the first query, but Grand Canyon National Park would not itself be flooded. The dams would not attack a sacrosanct principle or submerge a visual heritage. (The Colorado River had in fact been dammed by volcanic flows during the Pleistocene and had flooded the Canyon more than once; the soils farmed by the Havasupai were the deposits of one such episode.) The traditional conception of the Canyon as a scenic and scientific spectacle was an insufficient reply. The replumbed Colorado would not be visible from the rim, and Mission 66 would bring ever-larger numbers of tourists to not see those effects. If the projects were offensive or inappropriate, the reasons had to be found elsewhere.

Yet even as physical access improved, intellectual access declined. More cars brought more visitors to more overlooks, but it was not evident that they saw more deeply into the scene. If the Canyon was to endure as a culturally powerful presence, the public needed to find a meaning beyond souvenirs, hydropower, and scenic pullouts. A simple revival of Powell, Dutton, Holmes, and Moran was of limited use; they spoke the language of another age. The Canyon needed to express values deeper than pop culture and more contemporary than a recycled Romanticism.

* * *

The critical act of interpretation came when Joseph Wood Krutch published *Grand Canyon: Today and All Its Yesterdays* (1957). It was not a great book, not in the sense that the *Exploration of the Colorado River* and the *Tertiary History* had been, though the book had plenty of quotable passages. It was often sentimental and its science old-fashioned, premodernist in its simple evolutionism and its attachment to concepts like the Great Chain of Being. Its major audience had to wait until a few years afterward, when the book was converted into a successful television documentary. But with exquisite timing Krutch had written the prolegomenon to the conception of the Grand Canyon as wilderness.

Here was a bona fide man of letters, an éminence grise of high culture. Krutch, who had begun his career as a critic of books, plays, and society, was a man very much attuned to the spirit of the 1920s highbrow violently smitten by the village virus. In 1929 he consolidated a sophisticated case for intellectual despair with *The Modern Temper*, until the stock market crash pushed his book off a literary ledge and gave him genuine cause for grief. Even as sales fell, however, his reputation rose. Krutch settled into Columbia University and a role as a prominent drama critic. Gradually his strident, affected modernism mellowed, then vanished. After publishing a biography of Henry David Thoreau in 1948, Krutch made a Thoreauvian gesture and retired to open air and fresh thoughts in the Southwest. Now an apostate from modernism, Krutch recanted *The Modern Temper* with *The Measure of Man* (1954), a humanist paean; he began to write natural history books, in the spirit of Gilbert White and the reclusive Thoreau. Then he visited the Grand Canyon.

There the former spokesman for a world in which there was nothing to believe found a great constant, an eternal verity, and in writing of the Canyon, Krutch affirmed what had become for him a fundamental truth: "The wilderness and the idea of wilderness is

one of the permanent abodes of the human spirit." He successfully repositioned the Canyon within the moral geography of postwar America in which the wild and the synthetic, like two poles of
a bar magnet, defined the lines of force for understanding environmental issues. At the Canyon Krutch discovered a perfect expression in the contrast between the relict life of the Havasupai
Indians living in their side canyon Eden and the grotesque commercialism of Las Vegas, sprouting like a plastic mushroom in the
desert. Both were to compete for the waters of the Colorado; both
were to compete for the right to give meaning to the Canyon the
river had carved. One spoke with stone-inscribed truths; the
other, to a throwaway consumerism.[16]

The Canyon was known, but its significance had been relegated
to library shelves and its celebration to *Arizona Highways*; its
scenery was recognizable, and its visual effects seemed less stunning to a generation that had acquired television. Shrewdly
Krutch transcended utilitarian politics with an appeal to the wild
itself. The Grand Canyon was not hostile or even tangential to the
American Way of Life. Rather, it expressed the raw promise of
that new world and would continue to do so if Americans would
only leave it alone. "If we do not preserve it," Krutch concluded,
"then we shall have diminished by just that much the unique
privilege of being an American."[17]

Between them Stegner and Krutch redefined the Canyon's
meaning. They created a new poetry for the place and, unlike
those of the heroic age, did so on a basis other than science.
Though they wrote of scientists and alluded to scientific ideas,
they positioned the Canyon within a broad humanism; hagiogra-
phy and wilderness are not scientific concepts. It was well they did
so because modern science was engaged in its own developmental
schemes, the "endless frontier" proposed by Vannevar Bush and
funded by a host of military and civilian agencies, including the
freshly minted National Science Foundation. The old cultural alliance of the Second Age, like the Grand Alliance of the Second

World War, had disintegrated, leaving behind the intellectual iron curtain of C. P. Snow's "two cultures." Big Science now visited the Canyon as a tourist, not as a researcher.

Since the 1920s the earth sciences had been too often reduced to scholarly ritual and petroleum prospecting. Now that was changing. The same year that Krutch published his prose poem, on the centennial of the Ives Expedition, the International Geophysical Year (IGY) announced a Third Great Age of Discovery by launching Sputnik into earth orbit, lowering the bathyscape *Trieste* into the depths of the Marianas Trench, and dispatching a swarm of aircraft and tracked vehicles across the icy wastes of Antarctica. Geology at last roused from its long hibernation.

The Third Age took as its greater geographic domain the solar system. Its premier expression was the geophysical inventory of a planet, typically by remote sensing, a technique field-tested during IGY on planet Earth. The Earth-Moon system served the Third Age as the interior seas of Europe had the First Age and the natural history excursion and grand tour had the Second. Its defining gesture was a trek across the solar system, best epitomized by the two Voyager spacecraft. Its intellectual syndrome orbited eccentrically around modernism. The new explorations finally brought the revolution in earth science so long postponed during the twentieth century.

The Third Age resuscitated the American school. Its understanding of planet Earth culminated in the theory of plate tectonics. Comparative planetology, not the comparative frontiers of Western civilization, underwrote its prime insights. The Third Age did not even need human explorers, only robotic probes, and it was searching not for exotic peoples or lost civilizations but for evidence of life itself. Its moral drama was ambiguous. Contact with the terrains of the Third Age, often through prosthetic instruments, wiped away the moral crisis of prior exploring ventures but at the cost, it seemed, of moral drama and social significance.

If there was nothing to find, there was little reason to look. If there was little by way of new peoples and biotas to compromise exploration with empire, neither was there a compelling reason to proceed.

Of particular interest to the Grand Canyon was the character of geologic time. The new geophysical chronometers were not entropy clocks or life cycles. Like paleomagnetic reversals, or circadian rhythms, or the steady-state time of general systems theory, they were based on open rather than closed systems. The concept of information suggested that in such systems structures could maintain themselves, not necessarily decay. Meanwhile the development of modern genetics helped found evolution on chance rather than on design and helped destroy the value of the life cycle as a principle of metahistory. The world lines of quarks and tachyons replaced the depositional and diastrophic chronologies of epic earth history. Time was relativized; time's arrow, given a feedback loop. Colliding plates, unlike a contracting earth, suggested no necessary direction to geologic history. Impact craters, not canyons, recorded the early history of the solar system. Smashing moonlets explained the stratigraphy of the moon and Mercury. The Geological Survey's Branch of Astrogeology in Flagstaff scrambled down Meteor Crater rather than the Inner Gorge.

Its contact with new lands had meant the American school could claim a share of the national creation story: America as the frontier fusion of Old World civilization with New World wilderness. The vision of empty landscapes had encouraged an appeal for natural processes as geologic agents, working with infinite patience over aeons of time and a preference for long, exfoliating narratives. But the modern landscape was often rapidly resynthesized by human engineering, and it was necessary to understand and predict the almost instantaneous changes that might result. A sixty-million-year Tertiary history seemed arcane next to regional development schemes whose planning horizon extended over a decade. William Morris Davis's geographical cycle was powerless before the Upper Colorado River Storage Project. Geologic time compressed to a surface as nearly flat as LANDSAT imagery of the Canyon.

For the new earth sciences there was remarkably little of importance at the Grand Canyon. Their surveys migrated to the boundaries of plates, to the San Andreas fault, the East Pacific rise, the Hawaiian hot spot, the Mid-Atlantic Ridge; their ideas, to the macrofeatures of other Earthlike planets: the shield volcanoes and rift valleys of Mars, the highlands of Venus, the methane atmosphere of Titan, the seared dead surface of Mercury. They celebrated those macro features of the Earth that were comparable. The East African rift zone, the shield volcanoes of the Pacific, the channeled scablands of eastern Washington, the Sudbury crater in Ontario—here was the comparative geology that bound the Earth to the solar system. Vulcan's Throne from which Dutton and Holmes had peered into the western Canyon had to compete with volcanoes on Io and geysers on frigid Triton. Stratigraphers of plate tectonics looked to the Franciscan mélange in California's Coast Range rather than to the Redwall limestone. Where the dark schists of the Inner Gorge pointed to a Great Unknown, still older stromatolites from the West Australian craton spoke to the origin of life. In an era focused on gravity anomalies, paleomagnetism, heat flows, and continent-sized chunks of lithosphere that rammed into one another, the geomorphology that had once synthesized the American school—had traced the frontier between humanity and geology and between art and science—seemed merely decorative, like carved cornices in a steel skyscraper.

The Canyon no longer framed earth science as it had a century before. The Great Unconformity as testimony to the unbounded patience of geologic time faced the Chicxulub Crater, formed when an extraterrestrial body slammed into the Yucatán and precipitated the extinction of 90 percent of the planet's species, and made its far-scattered iridium a universal stratigraphic index. The slow recession of Canyon cliffs, testimony to the dynamics of the Great Denudation, had to match the immense storms of Jupiter's

Great Red Spot and Neptune's Great Dark Spot; the banded limestone, shale, and sandstone of Canyon buttes, the rings of Saturn and Uranus; the revelation of earth time with the cosmographic image of the Earth as a celestial body, like a sapphire set in black velvet.

On the centennial of John Wesley Powell's voyage in the *Emma Dean*, Neil Armstrong and Buzz Aldrin landed *Apollo 11* on the Sea of Tranquillity, and scientific eyes looked up, not down. In a memorial symposium on Powell, whose proceedings the Geological Survey published, Charles Hunt could ponder the evolution of the Colorado River system, noting that the fundamental questions of Canyon geology, including how it originated, remained unanswered. Luna Leopold, chief hydraulic engineer of the survey, replaced the evolutionary model of the Colorado with the proposal that it existed in a state of quasi equilibrium, transporting debris and water but no longer downcutting, and likened its rapids to thermodynamic engines. Mechanical metaphors seemed only appropriate when one was confronted with the geomorphology of wholesale engineering, when the floodgates of Glen Canyon had choked off spring floods, when the rhythm of the river beat to power demands created by regional air conditioners. The geologic puzzle endured as a historical curiosity, like unexplained artifacts at an Anasazi ruin.

That symposium built on one chaired by Edwin McKee in 1964. This session had attacked directly the question of an originating mechanism for the Canyon and retired in ordered ambiguity. Between those inconclusive symposia, meanwhile, a revitalized earth science had rapidly consolidated the formative concepts of plate tectonics. Even as the Powell symposium pondered the insoluble, a representative of the Museum of Northern Arizona—the institutional sponsor of the McKee symposium and a prominent promoter of Canyon science—discovered a fossil of *Lystrosaurus* that helped confirm the macromovement of continents. Ironically the find came not from the Colorado Plateau but from the mountains of Antarctica. The discovery by Edwin

Colbert was immediately hailed as major, confirmatory evidence to that already plucked from the deep ocean floor, traced along transcurrent faults, and mapped as volcanic arcs and overthrust zones where plates had subducted or overridden one another. The acceptance of plate theory further shifted attention to the boundaries of those lithospheric slabs, none of which framed the Grand Canyon. The Canyon resembled more a pothole on a plate's travels than the journey's end. The gorge ceased to be a scientific oracle.

The escalating Third Age challenged the Canyon's authority as a planetary spectacle. It revealed oceanic trenches like the Marianas, many times deeper; the sandswept Valle Marineris on Mars, many times wider and as long as the continental United States; ice-smothered basins in Antarctica, integral to the breakup of Gondwana. As an intellectual spectacle the geological aesthetics of Dutton had to confront the "geo-poetry" of Harry Hess, whose paper on mantle convection crystallized the revival of continental drift as a concept. The new imagery of the Canyon from orbiting spacecraft simplified its complexity and flattened its panorama of time; the Grand Canyon looked like a mud crack in the solar system. The organic ensemble that had made the Canyon so important a symbol to the American school now seemed as irrelevant as the Spencerian formula with which Clarence Dutton had described it, or the transcendentalism that animated Moran's paintings, or the sense of scientific revelation evident in the detail of Holmes's panoramas. The Canyon seemed to testify to a science as antiquated as trilobites.

The Third Great Age of Discovery would not revalorize the Grand Canyon.

It didn't need to. By an immense cultural metamorphosis the Canyon was being rediscovered. Tourists drove to the rim in renewed numbers and, more significantly, journeyed to and rode down the river at exponential rates. Even as Glen Canyon Dam throttled the Colorado's flow, the mainstream of American thought compensated and poured ideas through its gorges. Once again Americans were looking to nature for inspiration. It was not long before they looked to the Canyon.

By 1962 the two poles of postwar environmentalism each had a literary model: for the synthetic landscape, Rachel Carson's *Silent Spring*, which dramaticized the horrors of "chemical fallout"; and for wilderness, Eliot Porter's *"In Wildness Is the Preservation of the World"*, a blend of photography and prose. Even the National Park Service, still mesmerized by Mission 66, was ready for reform. Wallace Stegner meanwhile wrote an influential essay arguing for wild landscapes as part of "the geography of hope." Secretary of Interior Stewart Udall appointed a special committee to recommend better policies of wildlife management, particularly at Yellowstone. Chaired by A. Starker Leopold, a professor of wildlife biology at the University of California-Berkeley, a son of Aldo Leopold, the great philosopher of wilderness, and a brother of Luna Leopold, soon to begin the final hydraulic survey of the once-wild Colorado, the Leopold Committee quickly expanded its charge into a review of how the Park Service administered natural areas. In 1963 the committee's report recommended that parks should represent "vignettes of Primitive America." The next year Congress passed the Wilderness Act, which gave statutory standing to the concept of the wild and committed lands to its realization.[18]

Competition over the park's meaning, and controversy over its political future, heated up. The Bureau of Reclamation's Upper Colorado River Storage Project scheme was so enormous it demanded the two Canyon dams solely to sell hydropower to help underwrite its costs. This time there were two scenarios possible, symbolized by two books. If *This Is Dinosaur* memorialized a success story, its Sierra Club successor, *The Place No One Knew: Glen Canyon on the Colorado* (1963) documented a failure, the place the compromised Echo Park Dam had gone to. The choice seemed agonizingly clear. Grand Canyon could recapitulate either Echo

Park or Glen Canyon. That waters were already backing up behind Glen Canyon Dam—water that would otherwise flow through the sanctum of the Inner Gorge—gave urgency to the threat. The Grand Canyon dam controversy, as Roderick Nash observed, became "one of the classic confrontations in the history of American conservation."[19]

Glen Canyon had been flooded because its sandstone gorges had insufficient connections to American culture. Echo Park had been spared because a cultural elite had quickly found, or where necessary created, such connections. Meaning already resided in the Canyon, but the significance of the Canyon as scenic spectacle and as paean to earth time was insufficient by itself as counterarguments. The dams need not be seen from the principal overlooks; the contemplation of geologic time was lost on a throwaway culture that was committed to instant gratification and for which fifteen minutes of Warholian fame summed a celebrity lifetime. Public media were more interested in astronauts hiking down the Bright Angel Trail, promising to plant humanity's footprints on the moon, than on abstract arguments for the erasure of the human presence, hardly a telegenic subject.

The Canyon needed a cultural counterforce of equal magnitude and found it in wilderness. Like an old building gutted and rebuilt with a new interior, the cultural Canyon presented its classic facade while being internally rewired with the politics of ecology and reconstructed with more modern furnishings from the wilderness movement. Attention refocused on the river, as tourism did. Coincident with the Wilderness Act, the Sierra Club published a Grand Canyon book in the style of the Glen Canyon and Echo Park predecessors, François Leydet's *Time and the River Flowing: Grand Canyon*, and produced some films. While Leydet and company popularized the Canyon by boat, Colin Fletcher did it by backpack. Fletcher made his trip in 1963, published in 1967, and echoed the themes of Leydet et al.: the river and the awesomeness of earth time. But it was not really the Canyon as exhibitor of earth history that attracted them or drew a large audience. It

CANYON AND COSMOS / 153

was the Colorado as emblem of a wild and free-flowing river and the Canyon as timeless symbol of untrammeled nature, of rhythms longer than human experience and deeper than human knowledge.

The Canyon shone as park, as wilderness, as symbol of environmentalism's resolve. The court battles that followed the initial preservation of the Canyon had centered on the Antiquities Act and on the value of the Canyon for history and science; the new political fights were to focus on the meaning of wilderness. As Congress scheduled hearings in the summer of 1966 on the Pacific Southwest Water Plan, which was to oversee development on the lower Colorado, the Sierra Club sponsored a series of full-page ads in the *New York Times* and *Washington Post* in protest. The first argued that "there is only one simple, incredible issue here: this time it's the Grand Canyon they want to flood. *The Grand Canyon.*" A July ad mocked the Bureau of Reclamation's arguments that reservoirs would open the Canyon to public appreciation and shamed a constituency committed to conservation. "SHOULD WE ALSO FLOOD THE SISTINE CHAPEL SO TOURISTS CAN GET NEARER THE CEILING?" That put the matter to the public. David Brower, head of the Sierra Club and a veteran of Echo Park, put the matter to environmentalists. "If we can't save the Grand Canyon," he asked, "what the hell can we save?"[20]

The campaign succeeded, although the controversy did not cease until 1968, when Congress approved the Central Arizona Project without the offending dams, and also legislated a National Wild and Scenic Rivers System to expand the dominion of America's white-water wilderness of which the Canyon had become the paradigm. Simultaneously, now that Mission 66 had concluded, the National Park Service reformed its own house with a reconsidered set of guidelines for the management of the units under its care. The parks had a new meaning.

So did the Grand Canyon. What the dam episode symbolizes, and in some degree helped formulate, and what should not be obscured by the activist environmentalism, the politicization of

"ecology," and all the consciousness-raising that went on are that
the Canyon had reconnected with educated elites. Ideas came
from intellectuals who had read not only Dutton and Powell but
John Muir and Henry Thoreau and especially Aldo Leopold and
who saw wilderness as a great cultural resource, intimately bound
up with the American creation myth and American exceptionalism and with the human stewardship of nature everywhere. For all
its popular, sometimes mindless permutations, the cult of wilderness had real ideas behind it, and thanks to the dam controversy,
the Canyon became a repository for them. Wilderness advocates
had needed a symbol the public would recognize; the Canyon
needed a significance grander than scenery.

Like a broken bone, the fracture between the Canyon and intellectual culture knit together, stronger than ever. The Grand
Canyon became, for postwar environmentalism, both talisman
and oracle. It would again inspire as well as inform. Between a
white-water Grand and a dam-chocked Glen, intellectual energy
sparked, like an electrical arc leaping between oppositely charged
diodes. Edward Abbey set his novel of ecotage, *The Monkey
Wrench Gang* (1974), exactly within that force field.

In this revival the works of the heroic age were commonly quoted,
those from the Canyon's silver age hardly at all. With the exception of a few choice passages from J. B. Priestley, from the era between Dutton and Stegner there is barely a whisper. It hardly
mattered.

The renaissance rippled widely outward. Science returned; art
reappeared; literature renewed; the Canyon was rediscovered,
revisited, redefined. The Geological Survey restudied the river
preparatory to its symposium on the occasions of the Powell
centennial. The Museum of Northern Arizona and the Grand
Canyon Natural History Association published an anthology summarizing the state of Canyon geologic studies in 1974. Particularly

significant was a scientific transfer from geology to ecology as the dam-restructured river prompted a biological survey of the gorge. Separately the Geological Survey and the National Geographic Society remapped the Canyon. The National Endowment for the Arts began an artists-in-residence program in 1972, coinciding with a popular revival of western American subjects and representationalism as a genre; an Arts for the Parks program, under the aegis of the National Parks Association, superseded it. Still, landscape photography in the Eliot Porter and David Muench mode remained the principal form of Canyon art, finding ready patrons in the environmental periodicals or calendars, *Arizona Highways* magazine, and innumerable popular books. In 1983 a consortium of cultural institutions launched an annual Grand Canyon Chamber Music Festival. Within the long current of Canyon history, this brief era was a kind of Lava Falls, books foaming out of presses like white water against rock, as ideas met politics.

The river in fact became the Canyon's central metaphor. Because the park had been threatened by dams, the Canyon became equated with the Colorado River. The fate of one would be that of the other. The perspective of the river determined almost wholly the interpretation of the Canyon. Before the dam controversy, few tourists had traversed the gorge. Commercial boat trips, interrupted briefly by the impounding of Lake Powell, began to increase exponentially as controlled releases restored flow. Before 1963, when Glen Canyon Dam closed, fewer than 100 people had boated through Grand Canyon. By 1967 some 2,000 tourists were on the river, and by 1972, 16,400; the National Park Service found it necessary to regulate access with permits. The survival of Grand Canyon as a wilderness, it was argued, depended on the continuity of the river, which followed from the mechanics of boat travel. The Canyon's Colorado thus began at Lee's Ferry and ended in Lake Mead. Nothing should interrupt its free flow.

Virtually every work of significance described the integrity of the Canyon in terms of the unity of the river. For this too Echo

Park had established a precedent, as celebrity boat trips helped advertise the area. For Grand Canyon, Leydet dismissed the view from the rim as that of "an outsider looking in." Only by journeying "can you fully appreciate the work of the two great architects of the Grand Canyon, time and the flowing river," an observation that would surely have astonished Holmes and Moran, no less than Dutton and Powell, both of whom ascribed the splendor of the Canyon to the height and complexity of the Kaibab Plateau and found the relationship of rim to river the fundamental question of Canyon geology. "You sense it from an airplane," wrote Roderick Nash, author of *Wilderness and the American Mind*, a history of the idea of wilderness first published in 1967, but "you know it running the river." To the views of the writer and historian were added those of the artist. "You cannot grasp the scope of the Grand Canyon from the rim, however you try," explained Philip Hyde, photographer. "It is at the meeting of the river and the rock that those little things happen that make the landscape have meaning and sympathetic scale." It was an orientation that the Powell centennial did nothing to dispel.[21]

When the Park Service underwrote scientific work on the Canyon, as it attempted to do in the early 1970s, the emphasis was on the river. Even as recreational vehicles brought increasing throngs to the Canyon rim, a surge like those the Model T and railroad had swept up in earlier times, the Canyon had come to mean the river; the river, wilderness; and wilderness, ecology. In the nineteenth and twentieth centuries, geology and the institutions of geology had dominated conservation; now that mantle descended on biology. The river, curiously, had to be "alive," and the Canyon "living." The emphasis on the river was natural enough, not lessened by the arguments advanced on behalf of proponents of the dams that the water would not be visible from the eastern Grand Canyon viewpoints. It was natural to appeal to the river in order to define the Canyon as both a physical unit and a concept. But the equation with life was something new. To the uninitiated, political ecology might seem as curious a conceit with which to inform the Grand Canyon as Dutton's appeal to Spencerian evolution a century before.

Yet the propagandists and the intellectuals and even the public who argued that the Colorado River was Grand Canyon were on to something. As the Third Great Age of Discovery sent back its images and inventories of other planets, the uniqueness of planet Earth became more apparent. There it spun, with its stunning blues and whites, alone in the void, the Earth as precious object. Those shining colors came from water, which the Earth, perhaps solely, possessed in abundance and then, no less miraculously, within a temperature range near water's triple point. Water clouds, water oceans, water ice caps, and, of course, rivers—all were unique to Earth.

As a natural spectacle the Grand Canyon would be hard pressed to compete with the image of mass-fractured Miranda, gas storms on Jupiter that flared for centuries, filigree rings around white Saturn; a commercial boat trip through Colorado rapids was hardly on the order of a voyage across interplanetary space. But the history revealed by the Canyon was Earth history, not merely a history of geologic time, and it was a history unlikely to be found elsewhere in the solar system. The physical history of the Earth was distinctive in being by and large a geology of water: a history of erosion by water, deposition by water, sculpture by water. The Grand Canyon with its dramatic juxtaposition of rim and river served not only as a national but a planetary monument to that fact.

In this new cosmos the Canyon claims standing not because of its size or antiquity but, as Dutton had insisted, by virtue of its ever-evolving ensemble and the ideas continually made available by which to interpret it. How these might now converge is not obvious, but then it was never certain that Dutton and his colleagues were going to succeed in synthesizing landscape art, geology, expansionist politics, Romantic history, social Darwinism, aesthetic philosophy, travel literature, and the raw landforms of the West a century ago. The Canyon is unique, idiographic, dynamic; so are its interpretations.

But the grand convergence has endured, equally as a place and as a perspective. The importance of the Canyon will likely outlive the parochial American idea of wilderness, designation as a world heritage site, and mass tourism. A place that can hold a score of Yosemite Valleys and in which Niagara Falls would vanish behind a butte, that could absorb the shock of American expansionism and democratic politics, that could transcend a century of intellectual inquiry, from Charles Darwin to Jacques Derrida has not exhausted its capacity to refract whatever light nature or humanity casts toward it, provided a suitable overlook exists from which to view it.

That of course is Dutton's Point, which will be visited time and again, and made and remade. The Canyon has something yet to say, even if each visitor hears only the echo of his or her own voice.

AFTERWORD:

A REVIEW FROM POINT SUBLIME

Each man sees himself in the Grand Canyon.
—CARL SANDBURG

In September 1996 the North Rim Longshots held a reunion. The climactic event for the fire crew was an expedition to Point Sublime. We crawled over washed-out fire roads in whatever vehicles we could scrape together, recapitulating the numberless times we had raced to Sublime for a final look or a leisurely lunch. I sat on the outcrop at the point, lunch in hand, old comrades to my sides, and looked out. Suddenly time vanished. There was nothing before us to suggest that anything—anything *at all*—had happened in the last twenty-five years. There was no means by which to date the putative passage of time. There were no decaying trees, no new sprouts, no since-sculpted buttes; there was, at high noon, not even the measured tread of shadows. The scene was compelling testimony to the power of the Canyon not to reveal time so much as to obliterate it through sheer immensity. Twenty-five years blanked off like a deleted file. We sank into reverie.

That is not an experience possible with the present study. Its history is, for me, only too obvious, and the marks of editorial time both too apparent and too cautious. No one could confuse this edition with its callow first draft.

The present study began as a graduate term paper and in 1974 metamorphosed into a report in lieu of a thesis for a Master of Arts degree in American civilization at the University of Texas at Austin. Anyone familiar with the works of William H. Goetzmann and John E. Sunder will recognize their presence in its prose, a professorial mix of shock and patience that seems to me peculiarly suited to the study of a place like Grand Canyon. The extended essay was never intended to be published, nor deserved to be. Now that it has found a suitable form, I am pleased to dedicate it to those who first allowed me to push my homemade craft into the stream.

It was at the recommendation of Robert C. Euler, however, that I began to prepare the manuscript for publication by the Grand Canyon Natural History Association. That such a distinguished scholar of the Canyon should show interest was extremely flattering, for at the time I was a seasonal employee, the foreman of the North Rim fire crew; I am grateful for his faith in the project. Consequently I undertook extensive revisions. The text was almost completely recast, new insights introduced, some passages borrowed from my dissertation (a biography of Grove Karl Gilbert), and a good bit of old prose and shaky editorializing expunged. By then I had some tangible support from an unexpected quarter and, with it, a clearer sense of an organizing principle.

In 1981 I received the Antarctic Fellowship from the National Endowment for the Humanities and, after leaving the North Rim for the last time, spent an austral summer on the Ice. The two projects share a common conceit, the idea (first proposed by William Goetzmann) of organizing history according to ages of discovery. To Goeztmann's argument for a second great age, I added a third, and to *Dutton's Point*, as an inquiry into an earth emblem of the second age, I projected *The Ice*, a meditation on an earth emblem of the third age, Antarctica. This strategy was not obvious to NEH. In the end we survived. I thank the agency also for its support and patience.

A transfer to Arizona State University's main campus, after a decade at ASU West, introduced me to graduate assistants, and to our mutual surprise, I assigned Tonia Horton to help with clerical matters and data collection and then to review the old manuscript critically, which she did with zeal and insight. She has my thanks. So has my daughter, Lydia, who assisted on an expedition to the Grand Canyon research library. At Grand Canyon, Sarah Stebbins and Colleen Hyde helped me navigate through research collections. Hal Rothman has earned my gratitude for shocking me out of my Canyon coma long enough to revisit a favorite place. Cliff Nelson allowed me to reconnect with an old friend, even if I did it clumsily as a taxpayer. Thanks to Gerry McCauley for finding the manuscript a good home, and to Wendy Wolf for making it welcome and for forcing me to see the Canyon for the first time all over again.

Not least, thanks go to Sonja, my wife. Our lives together began on the North Rim. The Canyon continues to hang on our walls, salt our speech, and fill our lengthening lives, and remind us that we have never really left Dutton's Point.

APPENDIX

The Grand Canyon: A Graphical Profile

How the Canyon Became Grand is an extended interpretive essay that assumes rather than argues a variety of theses. While the following figures will not prove those arguments, they offer visual and, in places, quantitative support. They help place select Canyon processes within Canyon history and Canyon history within a larger landscape of historical scholarship.

Discovery

Figure A-1 plots out the major events of Canyon history by using the Colorado River as a time line. European history begins in 1540 with Cárdenas, located here at Mile 0, Lee's Ferry. The time line ends at Pearce's Ferry, 178 miles and 450 years later.

Figure A-2 graphs European exploring expeditions by fifty-year periods. There is a good bit of arbitrariness in counting, but there is every reason to believe the general contour is accurate. The First Age quickly reaches equilibrium. In effect, trade replaces exploration, while the dynamic of discovery continues because new nations send out exploring parties to try to outflank the established powers. The Second Age shows clearly; the heroic age of Canyon exploration occupies nearly the dead center of the era. The Third Age is still ambiguous. One scenario holds that the numbers will continue to decline to something like those of the First Age. Another, shown here, suggests that there is a revival under way, though far from the scale of the Second Age. Not fully counted for the Third Age are deep-ocean exploring sorties, many still classified from their Cold War origins. What will distinguish the Third Age is its peculiar ahuman (and abiotic) domain and its cultural alliance with modernism, broadly conceived. Sources: J. N. L. Baker, *A History of Geographical Discovery and Exploration* (New York: Cooper Square Publishers, 1967); J. H. Parry, *The Discovery of the Sea* (Berkeley: University of California Press, 1981); and Alex Roland, ed., *A Spacefaring People: Perspectives on Early Spaceflight.* NASA SP-4405 (Washington, D.C.: Government Printing Office, 1985).

GEOLOGIC TIME

Figure A-3 plots over the past four hundred years the commonly accepted ages of the earth. The significance of the Canyon as a demonstration of geologic time is immediately apparent. The valorization of the Canyon occurred amid this debate over earth history. Adapted from Preston Cloud, *Oasis in Space* (New York: Norton, 1988).

THE TOURIST CANYON

Figure A-4 records the annual visitation to the Grand Canyon, as measured against visitation rates to the national park system overall. The vulnerability of the Canyon during the late 1950s is clear, not only in terms of elite culture but in sheer numbers of tourists.

Figure A-5 isolates the Colorado River as a tourist experience. Made possible by the closing of Glen Canyon Dam, river running explodes exponentially at precisely the time that the dam controversy occurs. Those dams would have made the "natural" trip through the Canyon impossible. The rafting trip, in fact, helped redefine the dimensions of the Canyon overall. Source: Grand Canyon National Park.

THE CULTURAL CANYON

Figure A-6 shows the output of Canyon-related publications. General population growth provides a standard against which to measure the vigor of Canyon scholarship. The collapse from 1930 to the mid-1960s is striking and helps explain why the Canyon could be considered during that time as a site for high dam construction. Data from Earle Spamer, *Bibliography of the Grand Canyon and the Lower Colorado River* (Grand Canyon Association, 1990 edition).

A GRAND GEOLOGY

Figure A-7 traces the publication of Canyon-related scientific literature. Although geology consistently dominates, the post–Glen Canyon Dam research made biology an important challenger for interest and funding. Data from Spamer, *Bibliography of the Grand Canyon and the Lower Colorado River* (1990).

Geology's depressed publication output from 1920 to the 1960s resulted from changes within the science as well as from its relationship to the Canyon. Figure A-8 plots the age of citations within the literature of American geology. As the science became moribund, it shifted more and more away from the field and into the library. Not until the Third Age does the tempo of the science quicken. Source: Henry Menard, *Science: Growth and Change* (Cambridge: Harvard University Press, 1974).

THE PAINTED CANYON

Figure A-9 traces the growth of Canyon paintings, as "published" in some form or another: within books, exhibition catalogs, gallery collections, etc. Undoubtedly the list is incomplete, particularly in more recent years, but almost certainly the general contour is correct. Several items are clear. First, the early works had an impact out of proportion to their numbers. Second, after the heroic age, art followed rather than led the valorization of the Canyon. Third, until recently, one artist has tended to dominate the Canyon scene, in both numbers and influence. Astonishingly, Thomas Moran's influence extended for over forty years. Sources various.

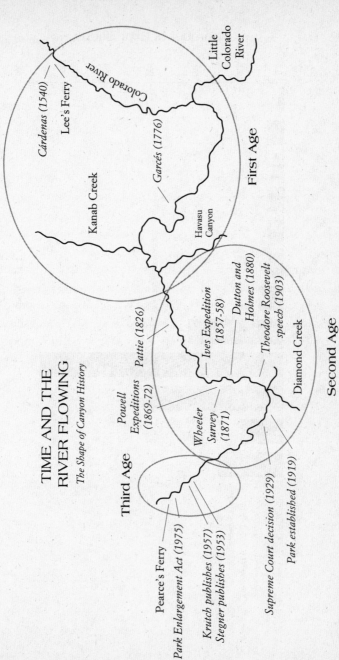

TIME AND THE
RIVER FLOWING
The Shape of Canyon History

First Age

Second Age

Third Age

Little
Colorado
River

Colorado River

Cárdenas (1540)
Lee's Ferry

Garcés (1776)

Kanab Creek

Havasu
Canyon

Pattie (1826)

*Powell
Expeditions
(1869-72)*

*Ives Expedition
(1857-58)*

*Dutton and
Holmes (1880)*

*Theodore Roosevelt
speech (1903)*

*Wheeler
Survey
(1871)*

Diamond Creek

Pearce's Ferry

Park Enlargement Act (1975)

Krutch publishes (1957)
Stegner publishes (1953)

Supreme Court decision (1929)

Park established (1919)

Figure A-1.

Second Great Age of Discovery

A Profile

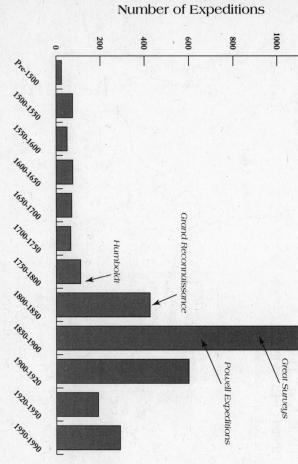

Number of Expeditions

Pre-1500
1500-1550
1550-1600
1600-1650
1650-1700
1700-1750
1750-1800
1800-1850
1850-1900
1900-1920
1920-1950
1950-1990

0   200   400   600   800   1000   1200

Humboldt

Grand Reconnaissance

Great Surveys

Powell Expeditions

Years

Figure A-2.

Data Sources:
J. N. L. Baker, *A History of Geographical*
  Discovery and Exploration
J. H. Parry, *The Discovery of the Sea*
Alex Roland, ed., *A Spacefaring People*

The Age of the Earth

History of an Idea

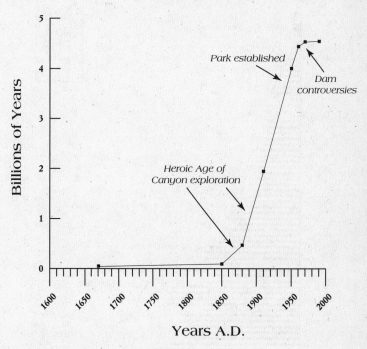

Figure A-3.

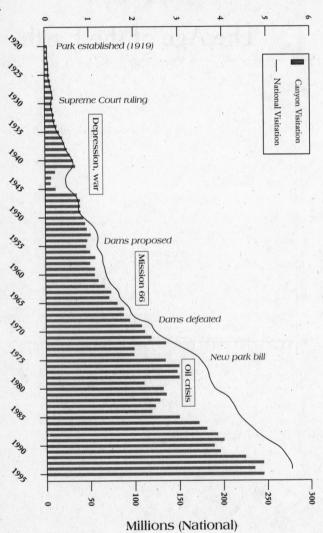

Figure A-4.

Grand Canyon

Contact and Visitation

Millions (Canyon)

Canyon Visitation
National Visitation

Park established (1919)

Supreme Court ruling

Depression, war

Dams proposed

Mission 66

Dams defeated

New park bill

Oil crisis

Millions (National)

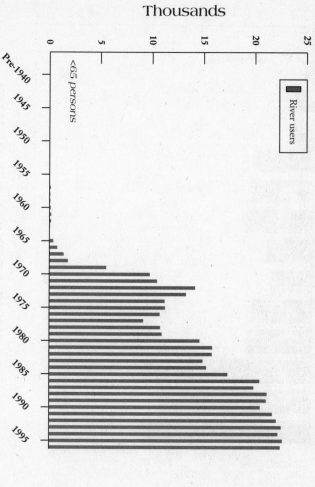

Down the River

Narratives of Rediscovery

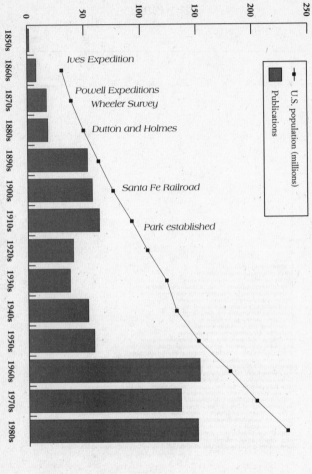

Grand Canyon Literature

General and Historical

Figure A-6.

Years

Ives Expedition

Powell Expeditions
Wheeler Survey

Dutton and Holmes

Santa Fe Railroad

Park established

■ U.S. population (millions)
▇ Publications

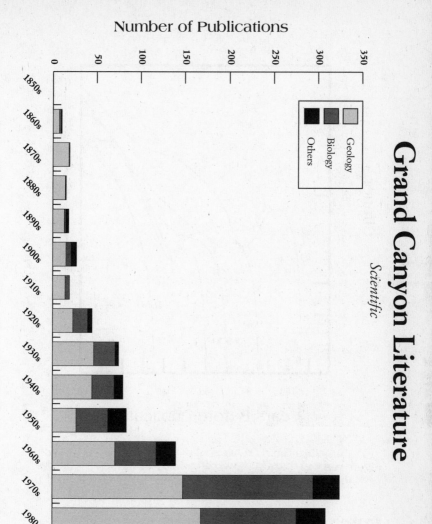

Figure A-7.

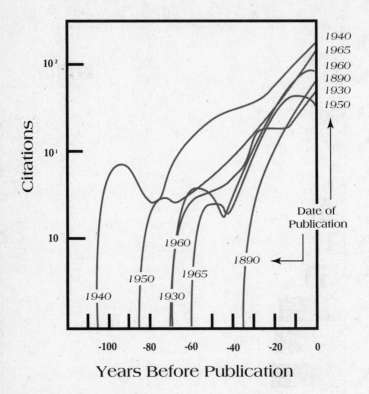

Figure A-8.

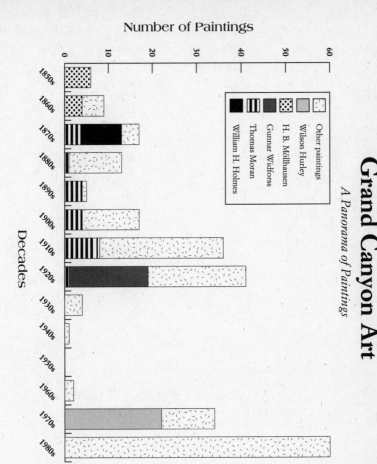

Figure A-9.

NOTES

Two New Worlds

1. Quoted from "Castañeda's History of the Expedition," p. 11, in Bruce Babbitt, ed., *Grand Canyon: An Anthology* (Flagstaff: Northland Press, 1978).
2. Ibid., p. 12.
3. See William Goetzmann, *New Lands, New Men: The United States and the Second Great Age of Discovery* (New York: Viking, 1986).
4. Elliott Coues, *On the Trail of a Spanish Pioneer*, vol. II (Francis P. Harper, 1900), pp. 347–50.
5. James Ohio Pattie, *Personal Narrative of James O. Pattie of Kentucky*, ed. Timothy Flint and Reuben Thwaites (Arthur H. Clark, 1905), p. 97.
6. Lieutenant Joseph C. Ives, *Report upon the Colorado River of the West*, Senate Ex. Doc., 36th Congress, 1st Session (Washington, D.C.: Government Printing Office, 1861), p. 21.
7. Humboldt, quoted in Douglas Botting, *Humboldt and the Cosmos* (New York: Harper and Row, 1973), p. 76.
8. See Edward S. Wallace, *The Great Reconnaissance: Soldiers, Artists, and Scientists on the Frontier, 1848–1861* (Boston: Little, Brown, 1955) and William Goetzmann, *Army Exploration in the American West, 1803–1863* (New Haven: Yale, 1959) and *Exploration and Empire* (New York: Knopf, 1966). Quote from Goetzmann, *Army Exploration*, p. 5.

RIM AND RIVER

1. Roosevelt quoted, Babbitt, *Grand Canyon*, pp. 187–88.
2. Ives, *Report*, p. 81.
3. Alonso Johnson quoted in Goetzmann, *Army Exploration*, p. 379.
4. See, for example, Ives, *Report*, p. 109.
5. Ibid., p. 110.
6. Humboldt quoted in Ben W. Huseman, *Wild River, Timeless Canyons: Balduin Möllhausen's Watercolors of the Colorado* (Fort Worth: Amon Carter Museum, 1995), p. 30.
7. Ibid., p. 68.
8. Ibid., p. 221.
9. Goetzmann makes this point twice, first in *Army Exploration in the American West, 1803–1861* (New Haven, CT: Yale University Press, 1959), p. 389, and again in *Exploration and Empire* (New York: Knopf, 1966), p. 330.
10. John S. Newberry, in Captain John Macomb, *Exploring Expedition from Santa Fe, New Mexico, to the Junction of the Grand and Green Rivers of the Great Colorado of the West* (Washington, D.C.: Government Printing Office, 1876), p. 54.
11. Dr. J. S. Newberry, "Geological Report," in Ives, *Report*, Part III, p. 45.
12. John S. Newberry, "Report of Progress in 1870," *Geological Survey of Ohio* (Columbus, Ohio: Columbus Printing Co., 1870), p. 14.
13. See Otis "Dock" R. Marston, "Who Named the Grand Canyon?," *Pacific Historian*, Vol. 12, no. 3 (Summer 1968), pp. 4–8; Samuel Bowles, *Our New West* (Hartford, Conn.: Hartford Publishing Co., 1869), p. 500; Powell quoted in William Culp Darrah, *Powell of the Colorado* (Princeton: Princeton University Press, 1962), p. 93.
14. John Wesley Powell, the *Exploration of the Colorado River of the West* (Cambridge: Harvard University Press, 1962, reprint), p. 83; Alexander von Humboldt, *Personal Narrative of a Journey to the Equinoctial Regions of the New Continent*, trans. Jason Wilson (New York: Penguin Books, 1995), p. 203.
15. J. W. Powell, *The Exploration of the Colorado River and Its Canyons*, (New York: Dover Publications, 1961; reprint of *Canyon of the Colorado* [Flood & Vincent, 1895], p. 275.
16. Powell, *Exploration* (Harvard edition), p. 93.
17. Powell, *Exploration* (Dover edition), pp. 390, 397.
18. Wallace Stegner makes the point nicely when he observes that no place, even a wild place, is a place until it has a poet. Certainly this was true for the Canyon and aptly describes Powell's contribution. Stegner, *Where the*

Bluebird Sings to the Lemonade Springs (New York: Penguin Books, 1992), p. 295.

19. Captain George M. Wheeler, "Geographical Report," *U.S. Geographical Survey West of the 100th Meridian* (Washington, D.C.: Government Printing Office, 1889), p. 170.

20. G. K. Gilbert to J. W. Powell, 09-24-75, "Letters Received," Powell Survey, Record Group 57, National Archives.

21. Captain G. M. Wheeler, *Report Upon United States Geographical Surveys West of the One Hundredth Meridian*, vol. I, Geographical Report (Washington, D.C.: Government Printing Office, 1889) p. 170.

22. Bailey Willis, *A Yanqui in Patagonia* (Palo Alto, Calif.: Stanford University Press, 1947), p. 33.

23. Personal telephone interview with Edwin McKee.

24. John Muir, "Wild Parks and Forest Reservations of the West," *Atlantic Monthly* (January 1898), p. 28.

25. Clarence E. Dutton, *Tertiary History of the Grand Cañon District*, U.S. Geological Survey Monograph 2 (Washington, D.C.: Government Printing Office, 1882), p. 141.

26. Wallace Stegner, *Beyond the Hundredth Meridian* (Boston: Houghton Mifflin, 1954), pp. 173–74.

27. The primary sources of biographical material on Dutton are: Wallace E. Stegner, "Clarence Edward Dutton: Geologist and Man of Letters," Ph.D. thesis (University of Iowa, 1935), later published as *Clarence Edward Dutton: An Appraisal* (Salt Lake City: University of Utah, 1936); George F. Becker, "Major C. E. Dutton [obituary]," *American Journal of Science*, 4th ser., vol. 33, no. 196 (1912), pp. 387–88; J. S. Diller, "Major Clarence Edward Dutton," *Bulletin of the Seismological Society of America*, vol. 1, no. 4 (1911), pp. 137–42, and "Memoir of Clarence Edward Dutton," *Geological Society of America Bulletin*, vol. 24 (1913), pp. 10–18; and an autobiographical sketch, dated December 20, 1886, contained in the Marcus Benjamin Papers, Record Unit 7085, Smithsonian Institution Archives, which is the major source of Dutton quotes in the referenced passage.

28. See autobiographical sketch in Marcus Benjamin Papers, loc. cit.

29. Clarence Dutton to J. W. Powell, September 17, 1876, "Letters Received," Powell Survey, Record Group 57, National Archives.

30. Clarence E. Dutton, "Mount Taylor and the Zuñi Plateau," U.S. Geological Survey, *Annual Report*, vol. 6 (Washington, D.C.: Government Printing Office, 1885), p. 113.

31. Clarence E. Dutton, *Report on the Geology of the High Plateaus, Utah* (Washington, D.C.: Government Printing Office, 1880), p. 284.

32. Dutton, *Tertiary History*, p. 141.

33. Ibid., p. xvi.

34. Ibid., pp. 142–43.

35. Samuel Emmons to G. F. Becker, June 16, 1882, "General Correspondence," George F. Becker Papers, Library of Congress.

36. Dutton quoted in Mary Rabbitt, *Minerals, Lands, and Geology for the Common Defense and General Welfare, vol. 2, 1879–1904* (Washington, D.C.: U.S. Geological Survey, 1980), p. 42.

37. Ibid., p. 104.

38. Dutton to W. H. Holmes, 1899, "Random Records . . . ," vol. 8, pp. 91–94, William H. Holmes Papers, National Portrait Gallery.

39. Dutton quoted by Diller, "Major Clarence Edward Dutton," p. 18.

40. Thomas Moran, "New World," in Charles Lummis, *Mesa, Cañon, and Pueblo* (New York: Century Co., 1925), p. 504. The exact date for Cole's *The Oxbow* is unclear. The best guess is that Cole worked on the painting between 1836 and 1838, with the latter date marking its completion.

41. Moran, quoted in G. W. Sheldon, *American Painters* (New York: D. Appleton and Co., 1879), p. 125.

42. Moran, ibid., p. 124.

43. Dutton, *Tertiary History*, p. 164.

44. Moran, "New World," p. 505

45. Moran quoted in Thurman Wilkens, *Thomas Moran: Artist of the Mountains* (Norman: University of Oklahoma Press, 1966), p. 4.

46. Quoted in Wilken, *Thomas Moran*, p. 238.

47. Dutton, *Tertiary History*, p. 142.

48. William H. Goetzmann, *William H. Holmes: Panoramic Art* (Fort Worth: Amon Carter Museum, 1977).

49. C. L. Walcott, "Pre-Carboniferous Strata in the Grand Canyon of the Colorado, Arizona," *American Journal of Science*, 3d ser., vol. 26 (1883), p. 437; Dutton letter to Geikie, quoted in Rabbitt, *Minerals, Lands*, p. 71. I have benefited from an unpublished paper, "C. D. Walcott and the Grand Canyon," that Ellis L. Yochelson, its author, generously shared with me.

50. William Morris Davis, "The Lessons of the Colorado Canyon," *American Geographical Society Bulletin*, vol. 41 (1909), p. 346.

51. Davis, "crack piece," quoted in Richard Chorley et al., *The History of the Study of Landforms; or, The Development of Geomorphology*, vol. 2, *The Life and Work of William Morris Davis* (London: Methuen, 1974), p. 727.

52. Geikie quoted in Gordon Davies, *Earth in Decay* (New York: Neal Watson, 1969), p. 352.

53. Roosevelt quoted in Paul Schullery, ed., *The Grand Canyon: Early*

Impressions (Boulder: Colorado Associated University Press, 1981), pp. 101–02.

CANYON AND COSMOS

1. James quoted in Richard Hofstadter, *Social Darwinism in American Thought* (Boston: Beacon Press, 1944), p. 129.
2. See Ferde Grofé, "Story of Grand Canyon Suite," *Arizona Highways*, vol. 14 (December 1938), pp. 6–9.
3. Bruce E. Babbitt, *Color and Light: The Southwest Canvases of Louis Akin* (Flagstaff: Northland Press, 1973), p. 23.
4. Katherine Plake Hough et al., *Carl Oscar Borg: A Niche in Time* (Palm Springs, Calif.: Palm Springs Desert Museum, 1990), p. 57.
5. Helen Laird, "Carl Oscar Borg," ibid., p. 43.
6. Bill Belknap and Frances Spencer Belknap, *Gunnar Widforss: Painter of the Grand Canyon* (Flagstaff: Northland Press, 1969), p. 25.
7. Ibid., pp. 74–75.
8. Archibald Geikie, "The Tertiary History of the Grand Cañon District," *Nature* (February 15, 1883), p. 357.
9. Ellsworth Kolb, *Through the Grand Canyon from Wyoming to Mexico* (New York: Macmillan, 1914), p. 3.
10. Ibid., p. 4.
11. Robert C. Euler, "Foreword," in Babbitt, *Grand Canyon,* p. ix.
12. J. B. Priestley, "Midnight on the Desert," ibid., pp. 103–04
13. For the expedition's story, see "Harold E. Anthony, "The Facts About Shiva," pp. 709–22, 766, and George B. Andrews, "Scaling Wotan's Throne," pp. 723–24, 766, *Natural History*, vol. XL, no. 5 (December 1937).
14. Priestley, "Midnight on the Desert," p. 104.
15. Samuel Trask Dana and Sally K. Fairfax, *Forest and Range Policy: Its Development in the United States*, 2d ed. (New York: McGraw-Hill Book Co., 1980), p. 426.
16. Joseph Wood Krutch, *Grand Canyon: Today and All Its Yesterdays* (New York: William Morrow and Co., 1957), p. 275.
17. Ibid., p. 276.
18. Stegner's quote from p. 197, as reproduced in Wallace Stegner, "The Meaning of Wilderness in American Civilization," pp. 192–97, in Roderick Nash, *The American Environment: Readings in the History of Conservation* (Menlo Park, Calif.: Addison-Wesley Publishing Co., 1968); A. S. Leopold et al., "Wildlife Management in the National Parks," *Compilation of the Adminis-*

trative Policies for the National Parks and National Monuments of Scientific Significance, rev. ed. (Washington, D.C.: Government Printing Office, 1970), p. 101.

19. Roderick Nash, "The Perils and Possibilities of a Park," in Roderick Nash, ed., *Grand Canyon of the Living Colorado* (New York: Sierra Club-Ballantine Books, 1970), p. 102.

20. Roderick Nash, *Wilderness and the American Mind*, 3d ed. (New Haven: Yale University Press, 1982), p. 230.

21. François Leydet, *Time and the River Flowing: Grand Canyon* (San Francisco: Sierra Club, 1964), p. 84; Nash, "Perils and Possibilities," p. 107; Hyde quoted in Robert Weinstein and Roger Olmstead, "Image Makers of the Colorado Canyons," *American West*, vol. IV, no. 2 (May 1967), p. 38.

SOURCES
AND FURTHER READINGS

The existence of Earle Spamer's magisterial bibliography of the Grand Canyon, regularly updated, makes it pointless for me to offer some perfunctory summary of sources. Besides, that kind of archival research was not my purpose in writing this book. The notes provide full citations for all the quotations lodged within the text. For any other literature connecting to the Canyon, consult Earle E. Spamer, compiler, *Bibliography of the Grand Canyon and the Lower Colorado River from 1540*, Monograph Number 8, Grand Canyon Natural History Association (1990).

Instead I would like to elaborate on those materials that helped me contextualize the Grand Canyon story and that might not turn up in conventional, place-specific bibliographies. Many are biographical studies of significant Canyon figures. Often these works were my point of departure into other realms of scholarship, and it should go without saying that I could have included hundreds of other studies. But these are the books with which I began and to which I continually returned.

EXPLORATION. Begin with William H. Goetzmann's classic trilogy: *Army Exploration in the American West, 1803–1863* (1959), which reaches a climax with the Ives Expedition; *Exploration and Empire: The Explorer and the Scientist in the Winning of the American West* (1966), which places Canyon discovery within the context of western American exploration; and *New Lands, New Men: The United States and the Second Great Age of Discovery* (1986), which situates American exploration within a global setting. Very useful too is William Goetzmann and Glyndwr Williams, *The Atlas of North American Exploration* (1992), which distills five hundred years of European discovery into a delightful suite of maps. Pair it with Carl I. Wheat, *Mapping the Trans-Mississippi West, 1540–1861,* 5 vols. (1957–63).

A useful summary of global surveys, despite its British bias, is J. N. L. Baker, *A History of Geographical Discovery and Exploration*, rev. ed. (1967). To complement it with a more visually graphic style, see Eric Newby, *The Rand McNally World Atlas of Exploration* (1975) and especially Glyn Williams and Felipe Fernandez Armesto, eds., *The Times Atlas of World Exploration* (1991). For understanding the character of the First Age, see the many marvelous works of J. H. Parry, but especially *The Establishment of the European Hegemony, 1415–1715* (1961), *The Discovery of the Sea* (1974), and *The Age of the Reconnaissance* (1963). For the ocean-based transition to the Second Age, see J. C. Beaglehole, *The Exploration of the Pacific,* 3d ed. (1966) and Jacques Brosse, *Great Voyages of Discovery: Circumnavigators and Scientists, 1764–1843* (1983). Goetzmann's *New Lands, New Men* (above) is the defining study of the Second Age. To appreciate New Spain's contributions, see Iris Engstrand, *Spanish Scientists in the New World: The Eighteenth-Century Expeditions* (1981). To understand Humboldt's status, see Helmut de Terra, *Humboldt: The Life and Times of Alexander von Humboldt, 1769–1959* (1955). And for John Wesley Powell's, see, in addition to Wallace Stegner's "biography of a career," *Beyond the Hundredth Meridian* (1953), William Culp Darrah, *Powell of the Colorado* (1950). For an introduction to the Third Age, see Stephen Pyne, *The Ice* (1986), and "Space: the Third Great Age of Discovery" in Martin Collins and Sylvia Fries, eds., *Space: Discovery and Exploration* (1994). Contemporary reviews of IGY are available in Sydney Chapman, *IGY: Year of Discovery* (1960) and J. Tuzo Wilson, *IGY: The Year of the New Moons* (1961).

EARTH SCIENCE. The historical literature on geology is large. William Sarjeant, *Geologists and the History of Geology: An International Bibliography from the Origins to 1978. Supplement, 1979–84* (1987), and the U.S. Geological Survey's *Bibliography of North American Geology* remain indispensable references.

Intellectual and institutional histories continue to thrive. Still useful is

Frank Dawson Adams, *The Birth and Development of the Geological Sciences* (1954). For physical geography, see Preston James, *All Possible Worlds: A History of Geographical Ideas* (1972). Of special significance to Canyon interpretation, however, are the three volumes by Richard Chorley et al., *The History of the Study of Landforms; or, The Development of Geomorphology* (1964–82). Volume 2 is an intellectual biography of William Morris Davis.

For American geology, George Merrill, *The First Hundred Years of American Geology* (1924) remains unsurpassed for breadth and detail. For an effort to tell the larger story through the career of one geologist, see Stephen Pyne, *Grove Karl Gilbert* (1980). On the U.S. Geological Survey, see Thomas Manning, *Government in Science: The U.S. Geological Survey* (1967) and Mary E. Rabbitt's multivolumed series, *Minerals, Lands and Geology for the Common Defense and General Welfare*, (1979–), and in distilled form, *The United States Geological Survey, 1879–1989* (1989). Overwritten but helpful as a complement to Goetzmann's *Exploration and Empire* is Richard Bartlett, *Great Surveys of the American West* (1962). For a comparison of themes and styles with the Third Age, see Don E. Wilhelm, *To a Rocky Moon: A Geologist's History of Lunar Exploration* (1993) and Henry Menard, *The Ocean of Truth: A Personal History of Global Tectonics* (1986).

Many critical Canyon scientists still lack bona fide biographies. In their absence, consider the following sketches: Ellis L. Yochelson, "Charles D. Walcott," *Geotimes* (March 1979), pp. 26–29, for a wonderful profile of an oft-overlooked individual; J. F. Kemp, "Memorial of J. S. Newberry," *Geological Society of America Bulletin*, vol. 4 (1893), pp. 393–406; Clifford M. Nelson, "William Henry Holmes: Beginning a Career in Art and Science," *Columbia Historical Society* 50 (1980), pp. 252–78; and Fritiof Fryxell, ed., *François Matthes and the Marks of Time* (1962). Ellis Yochelson has written a biography of Walcott's career in the Geological Survey, *Charles Doolittle Walcott, Paleontologist*, to be published by Kent State University Press.

The controversy over the age of the earth was so fundamental in defining geology that it is difficult to tease it out of general histories. A good introduction to the early eras is Charles Gillespie's classic *Genesis and Geology* (1951), while Martin J. S. Rudwick, *The Meaning of Fossils: Episodes in the History of Paleontology* (1985) helps move the story along. Perhaps the best source is the evolution of Arthur Holmes's *The Age of the Earth* (many editions). A survey of the debate as it existed at the time Dutton et al. were interpreting the Canyon is Joe Birchfield, *Lord Kelvin and the Age of the Earth* (1975).

The related controversy over Darwin shows few signs of abating; there is no point here in elaborating on that scholarly industry. As good a contextual summary as any is Ernst Mayr, *The Growth of Biological Thought* (1982).

ARTS. Western art, landscapes, and national traditions are the subject of innumerable studies. Three books that I have found notably useful are Barbara Novak, *Nature and Culture. American Landscape and Painting, 1825–1875* (1980), for its explication of trends that culminated in the heroic age of Canyon painting; Simon Schama, *Landscape and Memory* (1995), for showing how art contributes to a valorization of landscapes; and William H. Goetzmann and William N. Goetzmann, *The West of the Imagination* (1986), for elaborating the art criticism contained in the senior author's earlier *Exploration and Empire*. Several books illuminate the northern New Mexico school; consider, for example, Arrell Gibson, *The Santa Fe and Taos Colonies: Age of the Muses 1900–1942* (1983).

The Mesa Southwest Museum sponsored an exhibition on Canyon art in 1987, published as Tray C. Mead, ed., *Capturing the Canyon: Artists in the Grand Canyon* (1987). But most Canyon art remains within the context of biographical studies or retrospective exhibitions by individual artists. Published collections include: Bruce Babbitt, *Color and Light: The Southwest Canvases of Louis Akin* (1973); Helen Laird, *Carl Oscar Borg and the Magic Region* (1986); Katherine Plake Hough et al., *Carl Oscar Borg: A Niche in Time* (1990); Peter Hassrick and Ellen J. Landis, *Wilson Hurley: A Retrospective Exhibition* (1985); Bill Kelknap and Frances Spencer Belknap, *Gunnar Widforss: Painter of the Grand Canyon* (1969); and Donald J. Hagerty, *Beyond the Visible Landscape: The Art of Ed Mell* (1996). Uneven but enlightening is James D. Horan, *Timothy O'Sullivan: America's Forgotten Photographer* (1966).

Perhaps appropriately Thomas Moran has inspired a virtual cottage industry of studies and plentiful exhibitions. A good biography is Thurman Wilkins, *Thomas Moran: Artist of the Mountains* (1966). Recent commentaries include Nancy K. Anderson, *Thomas Moran* (1997).

ENVIRONMENTAL HISTORY AND CANYON HISTORY. For the reinterpretation of the Canyon as wilderness, there is no better beginning than Roderick Nash, *Wilderness and the American Mind*, 3d ed. (1983). The

Echo Park controversy that led to the Canyon dam controversy is the subject of several studies, especially Elmo Richardson, *Dams, Parks, and Politics: Resource Development and Preservation in the Truman-Eisenhower Era* (1973) and Mark Harvey, *A Symbol of Wilderness: Echo Park and the American Conservation Movement* (1994). A fascinating scientific critique from the same era is L. B. Leopold and T. Maddock, *The Flood Control Controversy* (1954). To see how the controversy merged with the general history of the park, see J. Donald Hughes's mysteriously named *In the House of Stone and Light* (1978). The political history is continued in Barbara J. Morehouse, *A Place Called Grand Canyon: Contested Geographies* (1996).

In recent years biographies have appeared for several major Canyon interpreters. Powell continues to fascinate (see Exploration). John D. Margolis, *Joseph Wood Krutch: A Writer's Life* (1980) places Krutch's Canyon book within his larger career, as Jackson Benson, *Wallace Stegner: His Life and Work* (1996) does for Stegner's Powell study.

INDEX

ILLUSTRATION CREDITS

FOR THE BEST IN PAPERBACKS, LOOK FOR THE

In every corner of the world, on every subject under the sun, Penguin represents quality and variety—the very best in publishing today.

For complete information about books available from Penguin—including Puffins, Penguin Classics, and Arkana—and how to order them, write to us at the appropriate address below. Please note that for copyright reasons the selection of books varies from country to country.

In the United Kingdom: Please write to *Dept. EP, Penguin Books Ltd, Bath Road, Harmondsworth, West Drayton, Middlesex UB7 0DA.*

In the United States: Please write to *Penguin Putnam Inc., P.O. Box 12289 Dept. B, Newark, New Jersey 07101-5289* or call 1-800-788-6262.

In Canada: Please write to *Penguin Books Canada Ltd, 10 Alcorn Avenue, Suite 300, Toronto, Ontario M4V 3B2.*

In Australia: Please write to *Penguin Books Australia Ltd, P.O. Box 257, Ringwood, Victoria 3134.*

In New Zealand: Please write to *Penguin Books (NZ) Ltd, Private Bag 102902, North Shore Mail Centre, Auckland 10.*

In India: Please write to *Penguin Books India Pvt Ltd, 11 Panchsheel Shopping Centre, Panchsheel Park, New Delhi 110 017.*

In the Netherlands: Please write to *Penguin Books Netherlands bv, Postbus 3507, NL-1001 AH Amsterdam.*

In Germany: Please write to *Penguin Books Deutschland GmbH, Metzlerstrasse 26, 60594 Frankfurt am Main.*

In Spain: Please write to *Penguin Books S. A., Bravo Murillo 19, 1° B, 28015 Madrid.*

In Italy: Please write to *Penguin Italia s.r.l., Via Benedetto Croce 2, 20094 Corsico, Milano.*

In France: Please write to *Penguin France, Le Carré Wilson, 62 rue Benjamin Baillaud, 31500 Toulouse.*

In Japan: Please write to *Penguin Books Japan Ltd, Kaneko Building, 2-3-25 Koraku, Bunkyo-Ku, Tokyo 112.*

In South Africa: Please write to *Penguin Books South Africa (Pty) Ltd, Private Bag X14, Parkview, 2122 Johannesburg.*